D1614203

The Artisan's Guide to
Crafting Distilled Spirits

The Artisan's Guide to

Crafting Distilled Spirits

Small-Scale Production of Brandies, Schnapps & Liquors

Bettina Malle & Helge Schmickl

Translated by Paul Lehmann

Spikehorn Press
Austin, Texas

The Artisan's Guide to Crafting Distilled Spirits

Warning: It is illegal to distill alcohol in the United States without first obtaining the proper licenses.

Spikehorn Press
4029 Guadalupe St.
Austin, Texas 78751 U.S.A.
512-220-0544
mailbox@spikehornpress.com • *www.spikehornpress.com*

Printed in China

Originally published as Schnapsbrennen als Hobby
©Verlag die Werkstatt, 2013
Front cover photography © dantemaisto/iStock/Thinkstock
Interior photograph credits © Thinkstock 81, 83, 85, 86, 89, 91, 94, 103, 104, 115, 116

Publisher's Cataloging-in-Publication
Malle, Bettina.
[Schnapsbrennen als Hobby. English]
The artisan's guide to crafting distilled spirits : small-scale production of brandies, schnapps & liquors / by Bettina Malle & Helge Schmickl ; translated by Paul Lehmann. — English-language edition.
pages cm
Includes index.
LCCN 2015933318
ISBN 978-1-943015-04-7
ISBN 978-1-943015-05-4 (ebook)
1. Liquors—Amateurs' manuals. 2. Distillation—Amateurs' manuals. I. Schmickl, Helge. II. Lehmann, Paul (Translator), translator. III. Translation of: Malle, Bettina. Schnapsbrennen als Hobby. IV. Title.

TP590.M19513 2015 641.87'4
QBI15-600050

Contents

Preface to the English-Language Edition

As fans of the original will notice, the first English-language edition of *The Artisan's Guide to Crafting Distilled Spirits* goes far beyond a literal translation of the German original. In addition to converting all International System of Units (SI) into U.S. customary units, we have expanded many sections with more detailed descriptions of methods and terms that, though originally comprehensible in German, due to a differing culture, varying practices, and diverging vegetations are occasionally not easily understood in the United States or other English-speaking countries. Furthermore, we have added broader descriptions in all cases of necessary accessories and additives to assist the reader in obtaining these articles outside of Germany and Austria. In a few cases, where a specific article is available in Germany only, you'll find listed a detailed source of supply.

We have been dealing with this subject since the early 1990s, and all that we have written about our methods has been checked and rechecked again many times. If you follow the instructions, especially concerning still constructions and distilling/fermenting processes, as accurately as possible, you will receive the promised top quality with your own products, and you will be rewarded with a broad variety of superior alcoholic beverages.

We wish you great success!
Dr. Bettina Malle and Dr. Helge Schmickl
www.schnapsbrennen.at

Preface to the First Edition

My father was an enthusiastic distiller, and he seems to have passed the hobby on to me when I was still in the cradle. As a young boy, I saw the colorful, bubbling liquids and how much my parents and their friends enjoyed them. It was many years before I realized why they were so interesting, but you probably know the reason yourself already. Perhaps this early exposure was also the reason why I ultimately decided to study chemistry? In any case, the hobby seems to be addictive, as I was also able to inspire an enthusiasm for it in my longtime partner, colleague, and wife, Bettina.

Our schnapps distilling seminars afford us the opportunity to get to know many hobbyist distillers and to learn about their desires and problems. This is what inspired us to write this book and satisfy these needs. It introduces the newest developments in the field, which we have "pilfered" to a large extent from our study of chemical engineering. It's a practical handbook that isn't intended to cover every bit of theory and data on distilling, but rather to show you how to actually do it yourself.

We wish to extend our gratitude to the participants in our seminars and the visitors to our website, who contributed significantly to this book through their numerous tips and suggestions.

The book is organized so that—after a brief introduction to the tradition of distilling and the definitions of a few key terms—you first learn how to make a mash. We start with the simplest method, where fruit is fermented "naturally." We then introduce methods and additives that can significantly improve these simple mashes and greatly increase your yields. High-quality distillates aren't all you can make from these mashes; you can also create delicious fruit wines.

The next chapter covers stills. Whether you're buying a still or constructing one on your own, this chapter will give you important guidelines to follow to avoid discovering later that your still doesn't work right at all or is constructed wrong. We also explain how stills work and how to use them.

The distillation process is covered next. In this chapter, you'll read about how to separate the heads, hearts, and tails in order to produce a great distillate. You'll also learn that you don't necessarily have to perform a double distillation. We also cover how to dilute the result to produce a drinkable product. Once you're familiar with the "art" of distilling, the next chapter gives you a variety of recipes for Obstlers (fruit brandy, consisting of a mixture of several different fruit mashes), grappa, and more. We provide you with a core recipe and invite you to imitate and experiment with it.

But you don't always have to produce schnapps from a mash: the next two chapters cover alcoholic infusions (fruit, herbs, or spices infused in tasteless, high-proof alcohol) and spirits (fruit, herbs, or spices placed in alcohol steam to extract their flavor). These methods are very simple and are sure to surprise you with their excellent results. You'll surely want to try some of the numerous recipes immediately.

The "essential oils" section covers the production of essential oils for use in fragrant oil burners, creams, perfumes, etc. This process is much easier than you might think!

Finally, we talk a little about the domestic and international drinking cultures, including how to properly bottle and label your results and the legal situation. We end the book with some questions we frequently receive and, of course, their answers.

We wish you the best of luck in your experimentation!
Dr. Bettina Malle and Dr. Helge Schmickl
www.schnapsbrennen.at

CHAPTER I

Foundations and Traditions of Distilling

The discovery of alcohol production took place approximately five thousand years ago. The exact time is impossible to determine, lost in the mists of prehistory, but it has been determined that the Egyptians were familiar with more than twenty-eight different types of wine by this point, most of which were made from grapes and dates. They also brewed beer from millet and wheat and fermented honey and hemp leaves into intoxicating beverages. Vinegar was also produced during this era. However, there is no evidence that distillation had been invented by this point.

It is presumed that arrack (brandy made from rice or molasses) was first distilled in India around 800 BC. Aristotle, the physician and natural researcher, was the first to describe the physical principles of distillation in the fourth century BC. He described the process of distilling seawater into drinkable water. At the time, sailors boiled sea water in their ships, and sponges or wool were submerged in the steam. Squeezing the condensed liquid out of these materials produced fresh water. However, there was not yet any practical application of this process to alcohol distillation in Europe.

Essential oils were probably already being produced by distillation in China, India, and Persia at the time of the birth of Christ. The Tatars of Mongolia and the Gobi Desert were also very innovative: they fermented mare's milk in leather hoses, which they then distilled into milk brandy. In Rome, Plinius enjoyed drinking hot wine, which was really just a precursor to modern *Glühwein* (German mulled wine). Unfortunately, no exact descriptions survive from this era because Emperor Diocletian ordered all records relating to alcoholic distillation destroyed.

When the Arabs conquered Spain in AD 700 and founded many schools and universities, they also brought distillation to Europe—they had previously invented distillation flasks and distilled alcohol. The Arabs used alcohol as a solvent for cosmetic purposes and called it *al-co-hue*, which means something like "finely ground shiny powder for eye makeup." The Russians and Poles made use of an interesting technique: they separated alcohol and water by freezing it!

▲ The art of distilling

Shortly thereafter, "burning water" or "aqua ardens" found its way even into Christian monasteries, and the first herbal liqueurs were made from herbs and alcohol. The Franciscan Ramon Llull and the alchemist and doctor Arnaldus de Villa Nova eventually helped "burning water" achieve widespread acceptance in Europe around AD 1280. They saw the philosopher's stone in the combination of water (wine) and fire. It is known that juniper berry distillates were used to treat plague victims in the fourteenth century.

One of the first books on the subject was the *Little Book of Distillation* by Hieronymus Brunschwig. This book contains a large number of historically valuable illustrations along with countless recipes.

▲ Ancient distillation, from Hieronymus Brunschwig's distillation book, Strasbourg, 1507

In the sixteenth century, Paracelsus finally gave the burning water its permanent name—alcohol—and wrote a detailed book on distillation. The rate of new developments quickly increased, leading to the modern distilling industry, which we know today through products like Scotch whisky, Bénédictine, calvados, vodka, Irish whiskey, eau-de-vie, Jenever, and cognac.

Until the 1970s and 1980s, schnapps had a rather negative reputation. It generally took the form of a high-proof liquor to put in your tea, and that was it. This was partially a result of the common methods of distillation used at the time, however. Especially in rural areas, the predominant fruits distilled were those that could not be used for anything else or that were no longer really fresh. This of course led to lower quality products. Over the last fifteen to twenty years, however, a clear change in this trend

Paracelsus ▸

has been noticeable. Many distilleries are making quality more of a priority, both in the fruit and in the distillation itself, because only the highest quality can produce a premium distillate. At the same time, schnapps has again become acceptable in polite society and is available in the best and most expensive restaurants.

Schnapps, spirituous beverages, distillates

Before we get started on the concrete schnapps production process, we need to briefly sidetrack to familiarize you with the basic terminology we'll be using.

The terms schnapps, spirituous beverages, distillates, etc., are general words for high-proof alcoholic beverages, and are too vague to give us any sort of information on how the drinks are produced.

The most important type of production is with a mash. It involves fermenting fruit, which forms alcohol. Distilled mashes are referred to as brandy or fruit brandy; this term also applies if the distillate is not stored in wooden casks. This is the only method in which alcohol is created: in all other methods, preexisting alcohol is used.

Distillation is a term that will come up often in the rest of the book. It refers to heating a liquid mixture or a liquid with dissolved substances to a boil, producing steam. The steam is then led through a cooling device, which turns it back into a liquid. The point of distilling a mixture is to concentrate the liquid with the lower boiling point (e.g., the alcohol in an alcohol-water mixture).

Overview of the different production methods ▾

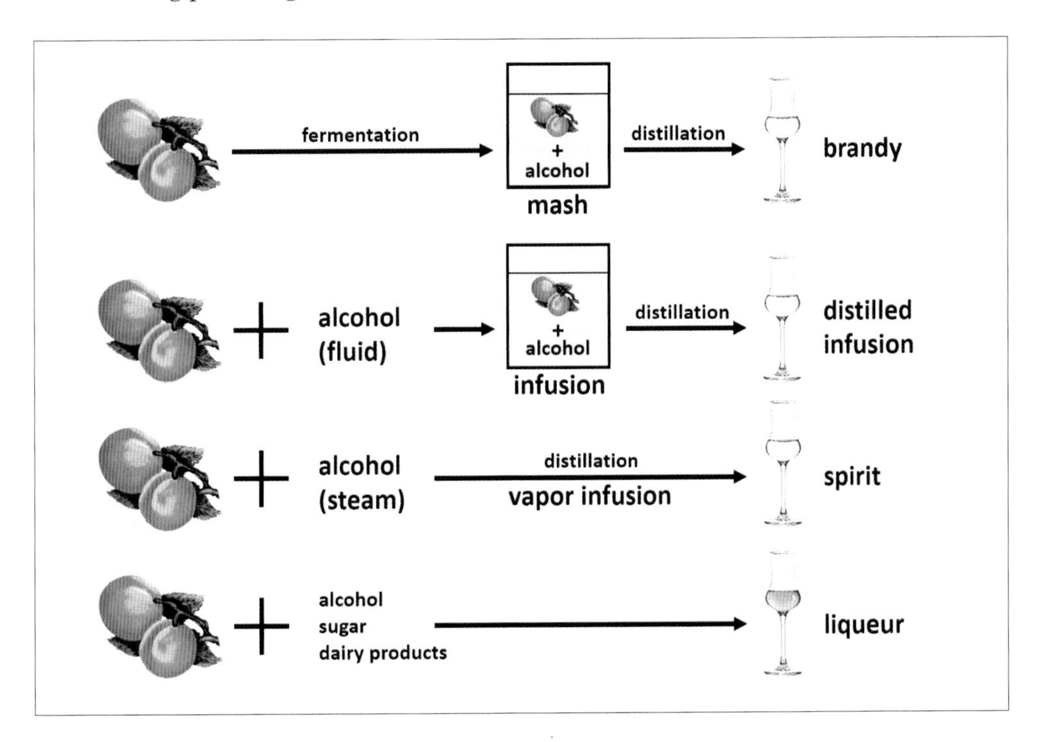

In other words, its purpose is to create a schnapps with a high alcohol content from a mash with a low alcohol content.

In an infused alcohol, fruits, herbs, etc., are pickled in a tasteless alcohol. The resulting infusion can then be distilled afterward, but it's not always necessary.

Finally, there is spirit production. In this method, the herbs or fruits are placed in the steam area of a still. If a tasteless alcohol is then distilled, it will take on the flavor of the fruits or herbs. This process is also called vapor infusion. No new alcohol is produced in this process.

It is also important to note that sugar is never added to schnapps in the German understanding of the term. Sweetened spirituous beverages are liqueurs. Milk products are often also mixed into them, giving them their usually creamy character.

CHAPTER 2

Mashes

Producing a mash is the first step toward creating your own alcohol. In the following chapter, we will address in detail exactly how this process works and what a hobbyist can do to attain the highest possible quality.

How is alcohol formed?

A side trip into chemistry

In chemistry, alcohol refers to a molecule containing at least one hydroxyl (OH) group. The class of alcohols therefore includes every compound containing this group, such as all types of sugar and glycerin, for example. In the formula, R denotes an organic substituent such as propyl or butyl.

R-OH

When speaking about alcohol without any further specification, so-called drinking alcohol is generally always meant. In chemistry, it's called ethyl alcohol or ethanol because it is composed of an ethyl group and a hydroxyl group. The ethyl group contains two carbon (C) and five hydrogen (H) atoms: CH_3-CH_2-OH.

CH_3-CH_2-OH

Ethanol has a boiling point of 173.3°F (78.5°C), which means that liquid alcohol boils at this temperature and turns to steam. In distillation, this steam is cooled back down (condensed) so that the alcohol again becomes a liquid.

Another well-known but not so well-loved alcohol is methyl alcohol, or methanol: CH_3-OH.

CH_3-OH

Methanol has a boiling point of 148.5°F (64.7°C), is extremely dangerous to your health, and can appear as a by-product of fermentation, for example if the mash contains many ligneous components such as stalks, leaves, etc.

Methanol should not be confused with heads (see chapter 4). If methanol has formed in the mash, it cannot be removed purely through distillation without special equipment.

The fermentation process

Fruit, berries, and herbs, the base substances of a mash, are made up of the following components:

- Water + sugar
- Flavors, aromas (essential oils)
- Vitamins, trace elements
- Solid, fibrous components (pits, skins)

Water + sugar

How can alcohol form from fruit? It's very simple: with yeast! Yeast consumes the sugar and then excretes alcohol. This process is known as "fermentation."

However, this does not work with solid, crystallized sugar or with solid piece of fruit, so the fruit must first be made into a pulp. If the fruit does not produce enough liquid, or if you're using herbs, roots, or blossoms, you will have to add water to it.

Yeast fungus is present in nature (on fruit), so simply leaving a fruit pulp standing for a couple of days may cause it to begin to ferment on its own (wild fermentation). That can happen, but may not, because there are also many other microorganisms in addition to wild yeast (e.g., rot, mold, etc.) that can decompose the pulp. Only yeast produces alcohol however; other organisms produce butyric acid, acetic acid, methanol, butanol, propanol, acetaldehyde, ethyl acetate, acetone, etc. Some of these substances are powerful toxins, so it's essential to keep out everything but the yeast fungus so that only the desired ethanol is produced. Since there is no way to keep these toxins from forming in wild fermentation, a fruit pulp should always be "seeded" with cultured yeast.

The *fermentation equation* shows how ethanol is formed from sugar (in this case, fructose):

$C_6H_{12}O_6$	$\xrightarrow{\text{Yeast}}$	**$2\,C_2H_5OH$**	**$+\ 2\,CO_2$**	**+ 88 kJ/mol**
Fructose	$\xrightarrow{\text{Yeast}}$	Ethanol	+ Carbon dioxide	+ Heat
35.3 oz. (1000 g)	$\xrightarrow{\text{Yeast}}$	18.0 oz. (511 g)	+ 17.3 oz. (489 g)	+ 488 kJ

The yeast breaks the sugar down into ethanol and carbon dioxide (CO_2), a reaction that also produces heat. The CO_2 is also why the contents of a fermentation barrel froth and bubble like champagne. Because of this, the barrel must not be sealed airtight without an air lock or it will burst!

Observe the ratio of the quantities of sugar (fructose) and ethanol. 35.3 ounces (1 kilogram) of sugar gives you 18.0 ounces (511 grams) of ethanol, which is a considerable amount, especially when you consider that this is 100 percent pure ethanol. If you reduce this concentration to a drinkable level (43 percent ABV), 35.3 ounces (1 kilogram) of fructose corresponds to 1.57 quarts (1.49 liters) of alcohol! A side effect of the fermentation is heat release, 488 kilojoules to be specific.

Flavors, aromas

Some people actually claim that fruit sugar tastes different from "normal" sugar and that fruits produce "better" alcohol than a pure water-sugar solution. This is simply not true. Both types of sugar, fructose (fruit sugar, contained in most types of fruit) and sucrose (cane or beet sugar, i.e., "normal" household sugar), are white, crystalline, completely taste-neutral substances. In their pure forms, both "normal" sugar and fruit sugar just taste sweet, but otherwise not like anything. It doesn't make any difference to the yeast either. Both types of sugar are broken down into the same substance, ethanol. There is no "better" or "worse" ethanol.

Independently of the sugar, fruit contains flavors and aromas (and essential oils in herbs, roots, etc.) that can be carried over into the schnapps through the mashing and distilling. The art of the mashing and the subsequent distilling is in retaining the taste and smell of as many of these substances as possible in the schnapps and not "losing" them. This can happen, for example, if volatile aromas are blown out along with the CO_2 during fermentation or if the aromatic substances are left behind in the kettle due to an improper distillation technique.

Vitamins, trace elements

If vitamins and trace elements aren't available to the yeast in sufficient quantities, deficiency symptoms may result, causing a reduced yield and undesired by-products. To avoid this, so-called yeast nutrients should always be added to the mash. Usually they contain diammonium phosphate, ammonium sulfate, trace minerals, and vitamin B_1. Add about 0.5 ounces (15 grams) per 25 gallons (100 liters) fruit mash. Malt extract and lemon juice can serve as a makeshift, if nothing else is available.

Solid, ligneous components

Substances such as methanol arise from solid, ligneous components. Branches, leaves, etc., therefore have no place in a mash, and ligneous or furry skins such as those of pomegranates and kiwis should also be removed. Pips are only a problem in excessive quantities. Fruits with cores (apples, pears, etc.) thus don't need their cores or their skins removed.

A high methanol content can be recognized by an unpleasantly sharp taste in the schnapps. For this reason, pomace schnapps tastes noticeably sharper than other schnapps varieties. Pomace is the leftovers from the presses used in wine production and is made up solely of crushed wine grape seeds, stems, and skins.

▲ Cherries

Which fruits can be used to make the mash?

In principle, any fruit, berry, or root can be fermented as long as they are not toxic. The following list provides a quick overview of the most common and popular bases for mash production:

- Pears
- Quinces
- Apples
- Raspberries
- Prune plums
- Currants
- Apricots
- Elderflowers
- Cherries
- Elderberries

Preparing the fruit

▲ Never use fruit that is rotten like this.

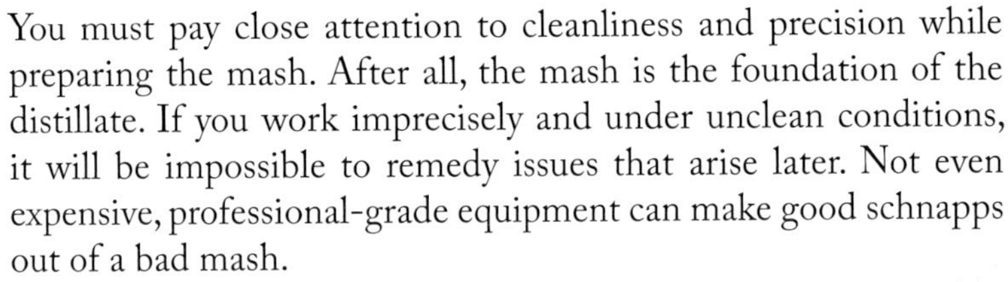

You must pay close attention to cleanliness and precision while preparing the mash. After all, the mash is the foundation of the distillate. If you work imprecisely and under unclean conditions, it will be impossible to remedy issues that arise later. Not even expensive, professional-grade equipment can make good schnapps out of a bad mash.

You should use the fruit as soon after harvesting it as possible, ideally on the very same day. If this isn't possible, you can freeze the fruit, which is the only way to keep ripe fruit fresh—otherwise, it will begin to accumulate rot and mold that cannot be removed later. It's best to use fully ripe "soft" fruit as these fruits are the most flavorful.

The first step in preparing the mash is to thoroughly wash the fruit. You can only skip this step in certain exceptional cases, such as if you're using elderflowers. Washing removes the small blossoms and pollen, so you should be sure to pick the flowers away from major streets, ideally a day after rainfall, to make up for the lack of washing.

Red currants are another exception, if they're picked without their stems. Washing causes a large amount of the juice to trickle out of the fruits, which is of course undesirable. Instead, pick the currants along with their stems and then wash them or pick clean berries without their stems shortly after it rains.

Note: What is the difference between fruits with pits (drupes) and fruits with pips (pomes)? Pits are always a single hard core, like in cherries. Pips, on the other hand, are numerous small seeds like you find in apples or pears.

During or after washing, you need to remove stems, leaves, and rotten areas from the fruit. If there are any completely rotten fruits, throw them away. It used to be common to make mashes from moldy fruit that had fallen to the ground or "not-so-nice" (i.e., actually just rotten) fruit, and to then make those mashes into schnapps. You absolutely must not follow this tradition! Rotten fruit, stems, and other impurities cause toxic heads to form. Putrefactive bacteria do also consume the sugar, but unlike with yeast, the result is not alcohol but rather the aforementioned organic solvents, some of which are toxic, which collect to a degree in the heads—but more on that in chapter 4. If you work under clean conditions, these problems will be kept to a minimum.

Once the fruit has been cleaned, you need to mash it up in order to make fermentation easier. Be careful when working with drupes (fruits with pits or stone fruit) such as plums, prune plums, mirabelle plums, apricots, cherries, etc.! The pits must not be damaged because doing so releases amygdalin, which produces toxic hydrogen cyanide. Because of this, some fruits like peaches or nectarines must have their pits removed prior to mashing because they break up relatively easily. Apricots should be mashed without their pits for taste-related reasons (see chapter 5). For all other types of fruits with pits, leave the undamaged pits in the mash and add about 10 percent of them to the kettle during distillation; otherwise the taste will be impaired.

Fruit pulp	Clean and wash the fruit, remove stems and rotten areas, and mash it. With drupes, make sure that no pits are destroyed.

You can check whether a mash contains hydrogen cyanide, cyanide, or carbamates using a testing set, for example the Cyan-EC-Test or the Schliessmann Cyanid-Test (successor of the Quantofix-Test), available in specialty shops. Experiments have shown that cyanide is only present to a measurable degree in distillates made from mashes that were produced with destroyed pits in them. Neither hydrogen cyanide nor cyanide could be measured in schnapps made from mashes with undamaged pits, regardless of whether they had undergone conventional or high-grade fermentation.

▲ Food mill, potato masher, juicer

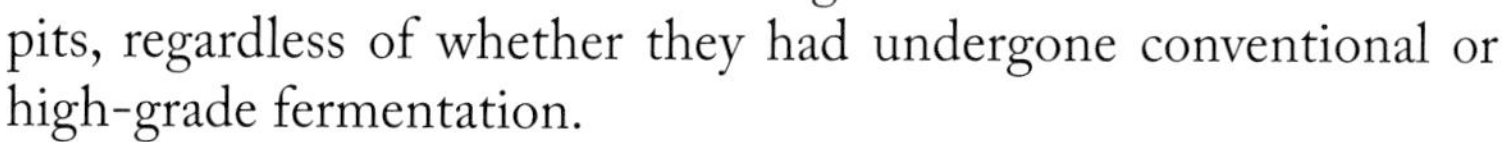

There are several different possible methods of mashing:

- Food mill, potato masher: this method works well if you're using a small quantity of soft berries.
- Juicer: ideal for hard tubers, potatoes, quinces, or carrots, but only in small quantities.
- Garden shredder, fruit grinder: appropriate for larger quantities of apples, pears, and especially quinces, but not for fruits with pits. They would destroy the pits. Clean the shredder thoroughly before you use it so that no residual wood from earlier makes it into the mash.

▲ Garden shredder

- Rubber boots + large tub: We've gotten our best results with this method. Fill a plastic tub to a level of 4–6 inches (10–15 centimeters) with fruit, then stomp the fruit with clean (!) rubber boots (making a mash can even contribute to your fitness!). This method avoids destroying the pits of plums, prune plums, mirabelle plums, or cherries, but it allows you to work with higher quantities (about 21 gallons or 80 liters of pulp per hour). We've also used this method with pears and apples because it's the easiest approach for batches of up to about 13 gallons (50 liters).
- Mixer attachment for power drills: This is the most effective method for working with soft fruits with pits (cherries, cherry plums, zibartes, etc.) because the hard pits won't break. It is also a good method for grapes (the berries only) if you don't have a winepress or for elderberries because their soft stems become entangled in the mixer attachment and can then be easily removed.

▲ The rubber boot method

Regardless of which method you use, or even if you come up with your own, the important thing is that the fruit flesh be turned into a pulp and, as already mentioned, that the pits not be damaged.

Fermentation container

Every food-grade plastic bucket or pail with an airtight lid or a 5- to 55-gallon barrel can be used as a fermentation container. The best materials are polyethylene (PE, HDPE) or polypropylene (PP). Do not use wooden barrels; they are difficult to clean properly, and microorganisms will develop in the tiny pores of the inner surface. If one is not already present, drill a hole in the lid for the fermentation lock. In some countries professional plastic fermentation barrels (also known as plastic fermenters, fermenting casks, or fermentation tanks), mostly equipped with a fermentation lock, are available in common hardware stores during harvest season. These barrels are offered in different sizes, starting at approximately 5–8 gallons (25–30 liters). Make sure the opening is wide enough for easy filling/emptying and especially for cleaning the barrel. Cider jars, made of glass or pottery, are possible, too, but the narrow bottleneck is inexpedient, especially when fermenting unfiltered chunky mash rather than juice.

▲ Plug and fermentation lock

What is the purpose of the fermentation lock, also called the air lock? As we already explained when describing the fermentation process, carbon dioxide gas (CO_2) is formed during the process. It must be able to escape from the barrel or else it will burst. No contamination should be able to get in, however, or faulty fermen-

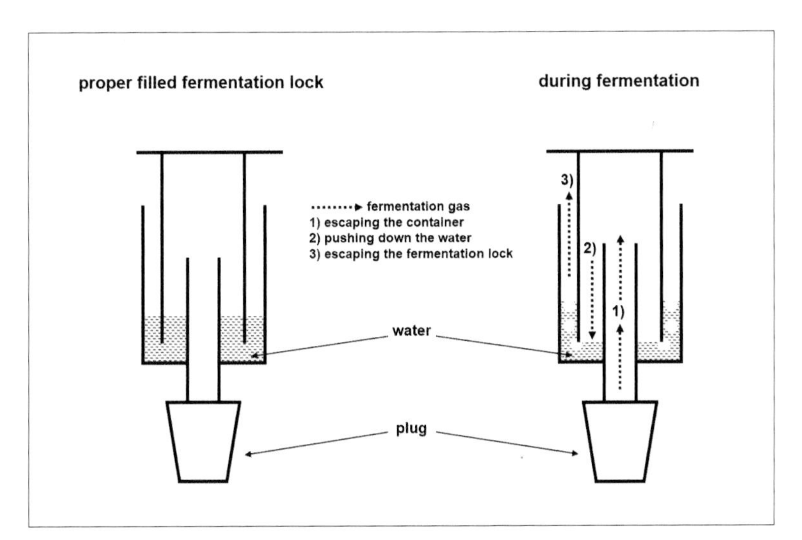

▲ Various different mash containers

◀ How the fermentation lock works

tation may cause vinegar or other products to form. It is especially important to mention fruit flies here. You have surely noticed that small insects suddenly appear whenever you leave fruit sitting in the kitchen for several days during the summer. These fruit flies carry bacteria that produce vinegar. If the flies and the bacteria they carry make it into the mash, a low-quality fruit vinegar may result, but of course your goal was to produce alcohol. Fruit flies seem to be able to smell a mash from miles away; in any case, we've never once performed a fermentation where these flies did not appear. A fermentation lock is therefore a very effective shield against vinegar formation.

Fill the fermentation lock up to the marked level with water. A small excess in pressure will now arise in the barrel, causing the gas to bubble through the water barrier, which lowers the pressure. This also keeps any air from entering the container through the fermentation lock, protecting the mash from outside bacteria.

☺ Tip: It shouldn't be a problem to construct a fermentation lock yourself using two old yogurt cups of different diameters and a pipe about 1–2 inches (3–5 centimeters) longer than the bigger cup. Drill a hole at the center of the bigger cup bottom and insert the pipe, leaving about 2 inches sticking out of the bottom. Seal the pipe with silicone caulk, fill water in the cup to about one-quarter of its height, and cover the construction with the upside-down smaller cup. The lowest part of the smaller cup has to dunk in the water.

Aside from the fermentation lock, the barrel must be airtight, so you'll need a seal between the cover and the barrel. If fermentation gas escapes through the cracks, this won't cause unwanted bacteria to enter the container, but it will keep the fermentation lock from performing its very important function of allowing you to check whether fermentation is proceeding normally. All you have to do is to check whether the fermentation lock is bubbling or not. If so, fermentation is proceeding properly. The fermentation will gradually slow down, so don't worry if bubbling frequency decreases after several days; that's normal behavior. You should always make sure that the fermentation lock isn't clogged; otherwise an immense pressure will build up in the barrel, until the loud bang.

Adding water

Pour the fruit pulp into the fermenting container, rinse out the equipment that you used to mash the fruit, and add this water to the barrel so that none of the fruit pulp is lost. A maximum of one-third of the volume of the fruit pulp should be added in water (normal tap water/drinking water). It's not necessary to add any further water: doing so dilutes the flavor.

There are only a few exceptions where more water is needed, such as elderflowers. They do not contain any water of their own, so you should add about an equal amount of water as the volume you have of the elderflowers. We have clearly marked these exceptions in the recipes.

Preparing the mash base

Fruit pulp	Clean and wash the fruit, remove stems and rotten areas, mash it. With drupes, make sure that no pits are destroyed.
Water	As much water as is necessary to clean the equipment (at most a third of the fruit pulp). Exceptions are noted in the recipes.

Conventional mashes

Fruit pulp without additional ingredients

The easiest way to make a mash is to just leave the fruit pulp sitting in the fermentation container, often even without a fermentation lock. If you're lucky, wild yeast will be activated by the fruit and alcoholic fermentation will begin after a couple of days. This approach is unfortunately also widespread in cider or perry (pear cider) production.

The fruits and the air also contain a multitude of other microorganisms that can also break down the fruit sugar, but without producing alcohol. As we've mentioned, the bacteria carried by fruit flies, for example, form vinegar—appropriate given their alternative name of "vinegar flies." Some other bacteria, on the other hand, produce toxic, terrible-smelling butyric acid or ethyl acetate and acetaldehyde (both major components of heads). All of these organisms eagerly feast on fruit pulp. Sometimes the alcohol-producing yeast prevails and a distillable mash results. But it often does not work like this, and you end up with a useless mash. Results vary from year to year. Sometimes you get lucky, sometimes you don't. This is why you often hear things like "the home-distilled schnapps didn't work out this year; last year was a much better season" in rural areas. But this variance really has nothing to do with the season, but rather with the microorganisms and which ones "won the battle for the barrel" this year. It is impossible to reliably guarantee quality using this method.

Another significant drawback of this method is the diminished yield: the large number of bacteria that come into contact with the barrel results in the formation of many (toxic) by-products, causing a rather significant increase in the amount of heads. And the wild yeast also need a portion of the mash's sugar to build new cells during the early stages of fermentation, leaving that sugar unavailable for alcohol production.

All you need for "wild" fermentation is fruit, water, and a fermentation container (with an air lock). However, as already described, using this method is not advisable. Unfortunately, however, mashes are still sometimes made this way today due to ignorance. The resulting schnapps has a very unpleasant pungent taste, and the taint of vinegar is unmistakable even to the inexperienced and cannot be gotten rid of even if you properly separate the heads during distillation. Usually, though, the heads are insufficiently separated or not separated at all in schnapps of this variety, leaving up to 10 percent (!) heads in the final "double-distilled" product.

Simple mashes

Fruit pulp	Clean and wash the fruit, remove stems and rotten areas, and mash it. With drupes, make sure that no pits are destroyed.
Water	As much water as is necessary to clean the equipment (at most a third of the fruit pulp). Exceptions are noted in the recipes.
Fermentation	Let it ferment with a fermentation lock.

Note: If you've made a mash that has already started to ferment on its own, there's no point in adding cultured yeast afterward. They cannot live together with wild yeast.

Adding cultured yeast

In order to markedly reduce the amount of heads (i.e., faulty fermentation), and to increase your yield of alcohol, it's a good idea to add cultured yeast. It ensures that the correct yeast fungi dominate from the beginning, steering the fermentation in the desired direction. Just adding yeast isn't enough, however. The mash also requires the aforementioned yeast nutrients. The trace elements and vitamins contained within them are indispensable to a top-quality fermentation because the yeast, much like us humans, needs them to grow and prosper. There are different types of cultured yeasts:

- Liquid yeast: Pure yeast to which the yeast nutrients must be added. Since getting fermentation to start can be problematic when using this type of yeast, it's a good idea to make a yeast starter (see page 36).
- Dry yeast: Also pure yeast without yeast nutrients, usually for distilling purposes optimized *Saccharomyces cerevisiae* (brewer's yeast).
- Mixtures: "Finished mixtures" contain dry yeast, pectinase (see page 18), or nutrients.

It's best to determine the exact dosages of the yeast and nutrients from the information packaged with them because they can vary significantly. For liquid yeasts, which are usually wine yeasts, add about 2 fluid ounces (50 milliliters), depending on the product, per 1 gallon (40 liters) of mash. To avoid problems with starting the fermentation, do not add the yeast directly to the mash but make a yeast starter first (see page 36). For dry yeasts, again depending on the product, add about 5 ounces (140 grams) per 25 gallons (95 liters) of mash. These yeasts you can mix directly into the mash without any complications; there's no need to make a yeast starter. As aforementioned, most of the dry yeasts originate from brewer's yeast. This yeast is better suited for fruit mashes due to its ability to ferment more types of sugar (e.g., pentoses) than wine yeast. An example of a mixture of dry yeast and pectinase is Gärfix (can be translated as "fast fermenting") by Oestreich GmbH, of which you need about 3.5 ounces (100 grams) per 53–56 gallons (200–250 liters) of mash. Turbo yeasts (see page 23), a mixture of dry yeast and nutrients, can also be used in conventional mashes (without adding sugar). The dosage is 4 ounces (115 grams) per 26 gallons (100 liters) if added to a fruit pulp instead of sugar water.

☺ Tip: In theory, you can also use normal baker's yeast. However, you should expect a reduced yield because it is not specially formulated for alcohol production and therefore only has a limited alcohol tolerance.

Most cultured yeasts can only thrive at temperatures between 59°F and 81°F (15°C and 27°C). If the temperature is too high, the yeast will die off, and if it's too low, it will become inactive. At temperatures of 68°F (20°C) and higher, fermentation takes place very fast, which causes large amounts of fermentation gas to quickly accumulate. This can also cause flavor to be "expelled." The ideal room temperature for fermentation is thus 59°F–66°F (15°C–19°C).

The fermentation temperature is especially sensitive when you're working with apricots or cherries. If apricots are fermented at 73°F (23°C), for example, the resulting schnapps is nearly tasteless.

The following table gives you a brief summary of the steps for producing a mash with cultured yeast:

Mashes with cultured yeast

Fruit pulp	Clean and wash the fruit, remove stems and rotten areas, and mash it. With drupes, make sure that no pits are destroyed.
Water	Use as much water as is necessary to clean the equipment (at most a third of the fruit pulp). Exceptions are noted in the recipes.
Cultured yeast	Liquid yeast + nutrients according to the packaging *or* Dry yeast + nutrients according to the packaging *or* Mixtures: about 3.5 ounces (100 grams) per 53–56 gallons (200–250 liters) of mash
Fermentation	Let it ferment with a fermentation lock.
Note: If the mash is boiled before adding the yeast or if you use juice from a steam juicer, you should mix the mash or juice several times a day after the yeast addition, otherwise the oxygen content will be too low for yeast growth.	

Checking the pH value

Yeast growth is also dependent on the acidity (pH value) of the mash. If the pH value is too high or too low, the yeast will be destroyed or unwanted bacteria will be able to grow, leading to faulty

The pH scale can be divided into three ranges:

pH < 7: Acidic range (e.g., vinegar, citric acid, hydrochloric acid)

pH = 7: Neutral range (e.g., water)

pH > 7: Basic range (e.g., ammonia, sodium hydroxide solution, lime)

fermentation and the corresponding formation of toxic heads components.

A pH value of 3.0–3.5, which corresponds approximately to the acidity of fresh apple juice, is ideal for fermenting fruit and for the yeast required for the process. Because this level lies on the acidic end of the spectrum, maintaining it is sometimes called "acid protection." If the pH is not acidic enough, problems such as the aforementioned vinegar taint will appear in the distillate. The formation of a "yeast film," or mold film, is also caused among other things by too low acidity, thus a too-high pH value. Microorganisms (Mycoderma), basically a mixture of aerobic bacteria/mold/yeast, float on the surface and spoil the mash by oxidizing ethanol into acetaldehyde.

So how can we measure the pH value and correct it when necessary?

Measuring the pH value

The easiest and cheapest way to measure pH is with a pH strip. This is a small strip of cardboard or plastic prepared with an indicator solution. If the strip is inserted into a liquid, the indicator changes its color depending on acidity. The various colors and their corresponding acidities (pH values) are listed on the packaging. So just dip the strip in, check its color, and read off the pH value.

If the color change is imperceptible because of the color of the mash itself, such as when using blackberries, it's best to use "Biogen M" (a liquid acid blend, see next section) for acidifying because it adjusts the pH to the correct value by itself.

Another much more expensive and laborious method is a pH electrode. Even acquiring such an electrode is relatively expensive. If you dip the glass electrode into the liquid you wish to measure, its pH value is shown on a digital readout. You cannot simply put away the electrode after the measurement; after you use it, you have to clean it agonizingly thoroughly with distilled water and then store it in a special solution, which isn't exactly cheap. The electrode must be calibrated regularly, and if it dries out even once it will be ruined. This is really more something for chemists to bother with; the readily available indicator strips suit our purposes excellently.

Correcting the pH value

Other than a few exceptions, such as lemons, the pH of fruit mashes is generally too high, thus the acidity is too low. This means that, as a rule, the mash will need to be acidified. In case of lemon juice, add food-grade precipitated calcium carbonate or potassium carbonate to increase the pH. There are several different ways to acidify fruit mashes:

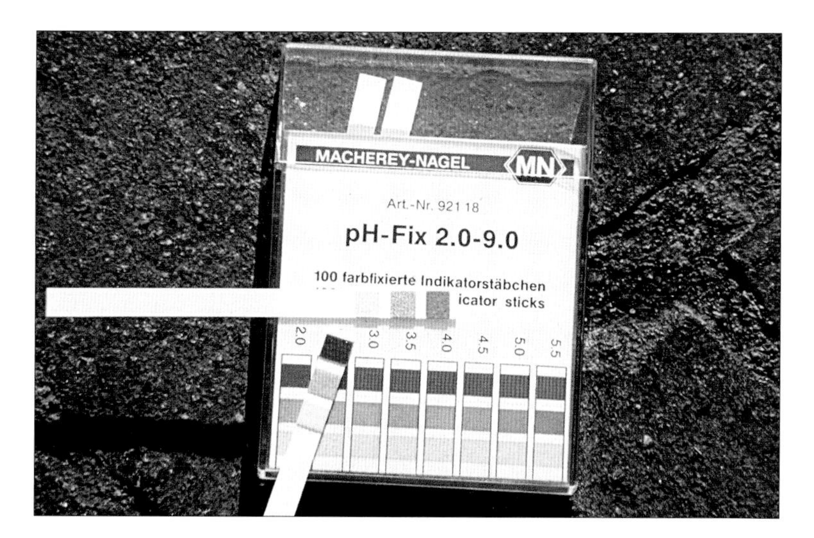

◂ pH strip: before the measurement, after the measurement, and a comparison chart

Lemons, oranges: Either squeeze them or add slices of untreated lemons/oranges to the mash, about three oranges and four lemons per 8 gallons (30 liters) of mash. This method only works, however, if their strong flavor won't interfere with the mash or will give it a desirable hint of flavor, like with elderflowers, for example. More on this in the recipes starting in chapter 5.

Citric acid: Cheap and available in any supermarket. However, due to the alcoholic fermentation, it breaks down after a few days. This means that you have to check the pH again every three or four days and correct the value again if necessary.

Instructions: First, determine the acidity of the mash using a pH strip. If it's too high, add one to two teaspoons of acid per quart (liter) of mash, stir it thoroughly, and measure it again. Repeat this process until the pH value is between 3.0 and 3.5. Unfortunately there is no simpler method with citric acid, because if you add too much, the pH can drop below 3.0, which will cause the yeast to die off because conditions are too acidic. It's also best to dissolve the citric acid in a little water before adding it to the mash. Otherwise it's possible for the powder to not yet have dissolved when you measure the pH.

Lactic acid: A big advantage of lactic acid compared to citric acid is that it does not break down during fermentation. This means that you only have to add it once, at the beginning of fermentation, and that no further corrections are necessary.

Instructions: As with citric acid, first measure the pH of the mash, and then add about a tablespoon of lactic acid per quart (liter), then measure it again. Repeat this process until you reach the proper pH value. You should also be careful in this case not to add too much to keep conditions from becoming too acidic for the yeast. Winemaking and distiller suppliers sell 80 percent solutions of lactic acid as well as complete sets for proper dosing. In this case, the mash must be titrated first (meaning slowly dropping a solution into a sample until there is a color change) and the results used to calculate the exact amount to be added. The pH strip method is entirely sufficient, however.

If you have added too much citric acid or lactic acid and brought the pH down below 3.0, add the aforementioned carbonates.

Acid blends as used in winemaking are a mixture of citric acid, malic acid, and tartaric acid; the proportion of the contents depends on the product. This type of wine additive is widespread and available at nearly every winemaking supplier. Although most acid blends are a powder, there are also fluid blends like Biogen M by Oestreich GmbH. In our experience, Biogen M is easiest to handle and measure because you cannot add too much: since a chemical buffer solution forms when it is combined with the fruit pulp, simply add the amount listed on the bottle, about 10 fluid ounces per 25 gallons (300 milliliters per 100 liters), to create the correct pH value. The pH will be at the right level (3.2–3.5) for the entire fermentation and beyond, regardless of what it was before the acidification.

Now that the mash's pH value is in the proper range, very little faulty fermentation should take place. You will only need to separate a very small amount of heads during the distillation. The table on page 19 gives the individual steps again.

Pectinase

When making a mash, you'll often encounter the problem of the fruit gelatinizing (i.e., the mash becoming gelatinous), making it almost impossible to work with any further. The yeast itself does produce an enzyme known as pectinase that helps inhibit this process, but it's often not enough. To deal with this issue, additional quantities of this enzyme, also known as "liquefier," are added. Again, this is not a foreign additive but rather a substance that the yeast produces itself, but in insufficient quantity. Another common name for pectinase is anti-gel.

The pectinase also helps ensure that the fruit flesh is more fully decomposed, transferring more of its flavors into the mash. The more liquefied the mash and the more the fruit pulp has dissolved, the better it will taste in the end. From the very first day after adding yeast, the pulp settles above the liquid—it is pushed upward by the fermentation gas. At the beginning, the liquid only reaches about one-tenth of the total fill level; the rest is more or less solid fruitcake. During fermentation, the solid parts disintegrate more and more to the point where the ratio is reversed by the end.

Commercial pectinase is derived from the fungus *Aspergillus niger*, used to clarify fruit juices (such as apple and grape) and for wine manufacturing. Most suppliers offer solid pectinase granulate, but liquid pectinases have extra-high activity, for example Enzeco Pectinase, Novozymes Pectinex Ultra Mash, Endozym Micro, RapidaseC80 WLN4800 by WhiteLabs, Verflüssiger Spezial (can be translated as "liquefier special") by Oestreich GmbH, etc. Despite their names, commercial products mostly also contain cellulase and hemicellulase.

Again, it's best to determine the exact dosage from the information in the packaging. About 1–4 teaspoons per 25 gallons (3–25 milliliters per 100 liters) of mash is a general reference point. When using a mixture of dry yeast and pectinase, you obviously don't need to add any extra.

With the pectinase, the perfect mash is now finished. As long as you've worked under clean conditions, there will be barely any heads to separate during the distillation.

Conventional mash with all necessary additives

Fruit pulp	Clean and wash the fruit, remove stems and rotten areas, and mash it. With drupes, make sure that no pits are destroyed.
Water	As much water as is necessary to clean the equipment (at most a third of the fruit pulp). Exceptions are noted in the recipes.
Cultured yeast	Liquid yeast + nutrients according to the packaging *or* Dry yeast + nutrients according to the packaging *or* Mixtures: about 3.5 ounces (100 grams) per 53–56 gallons (200–250 liters) of mash
Pectinase	Verflüssiger Spezial: 0.1–0.8* fluid ounces per 25 gallons (3–25* milliliters per 100 liters) of mash, if not already included in the yeast mixture

pH correction	Measure the pH value, correct it to 3.0–3.5 if necessary with Citric acid (adjust with a pH strip) *or* Lactic acid (adjust with a pH strip) *or* Acid blends: e.g., fluid blend Biogen M: 10 fluid ounces per 25 gallons (300 milliliters per 100 liters) of mash *or* Three oranges and four lemons per 8 gallons (30 liters) mash
Fermentation	Let it ferment with a fermentation lock. Optimal temperature: 61–66°F (16–19°C). Stir once or twice a week, stop stirring when the fruitcake has sunk; at that point the fermentation is finished.
Storage	Store the fully fermented mash at most one to two months; store brandy at least two, better three to four years for an edible result.
* The amount depends on the fruit. Soft fruit like drupes need less pectinase than hard fruit like quinces, rose hips, or rowanberries. Roots like Jerusalem artichokes need 0.8 fluid ounces per 25 gallons (25 milliliters per 100 liters).	

Alcohol content of conventional mashes

The following table gives you an overview of what alcohol contents various fruits can provide the mash with. It's impossible for more than that amount to form because the fruit's sugar will be used up at that point and the yeast will starve.

If the fruit you wish to use is not included in this list, you can estimate the maximum attainable alcohol content using the fruit's carbohydrate content, given in the nutritional information. A single ounce (1 gram) of carbohydrates corresponds to about 1 ounce (1 gram) of sugar. Example: 3.53 ounces (100 grams) of fresh pears contain 0.437 ounces (12.4 grams) of carbohydrates, so a quart (liter) of pear pulp has about 4.37 ounces (124 grams) of sugar. 4.37 ounces (124 grams) of sugar then corresponds to an attainable alcohol content of 6 percent ABV (see alcohol content/sugar table in the next chapter).

Attainable alcohol contents in conventional mashes

Alcohol content [% ABV]	Type of fruit
2–3	Juniper berries, rowanberries, raspberries, and other berries
3–4	Quinces, rose hips
4–5	Sloes, kiwis, papayas
5–6	Apples, pears, pineapples
6–8	Cherries, sour cherries, apricots, peaches, pomegranates, mangos
8–10	Plums, prune plums, mirabelle plums, bananas
10–13	Grapes

High-grade mashes with added sugar

The previous sections described how to produce a high-quality mash, but the quality can be even further improved and the alcohol yield greatly increased. This is only possible using a special turbo yeast or a sherry yeast. No other type of yeast will work because of their lower alcohol tolerance. They die off at a maximum ABV of 13 percent (wild yeast and baker's yeast at an even lower 6–8 percent) because higher alcohol contents are poisonous to them. However, the yeast alone is not enough. The sugar in the fruit itself is only sufficient to reach an alcohol content between 2 percent (using rowanberries) and about 8 percent (using apricots or prune plums). Grapes have the highest sugar content: dry, fully fermented wines have an ABV around 12–14 percent. To raise the alcohol content higher, the yeast must be supplied with additional nutrients in the form of sugar (refined sugar, such as normal household sugar), which the yeast then consume and turn into alcohol.

As we've already described in detail, the same chemical substance (ethyl alcohol or ethanol) is produced regardless of whether the sugar derives from the fruit or is added separately. The claim that you can get a headache from drinking schnapps where sugar was added to the mash is false and utter nonsense. Headaches are caused by the by-products of fermenting an unclean mash.

Note: If you're experienced with conventional mashes and distilling them, and you're satisfied with the quality of your distillates, you should try the high-grade method with a small amount. You may be surprised by the result! In German we say: "The enemy of the good one is the better one."

What are the advantages of a high-grade mash?

- The alcohol content in the mash can reach 20 percent ABV when using turbo yeast and 16 percent ABV when using sherry yeast. The higher alcohol content extracts or "pulls out" more of the flavor from the fruit than in conventional mashes because ethanol is an organic solvent and is therefore more similar to oil or fat in terms of its characteristics than to water. The mash thus takes on a much more intense flavor because flavorful and aromatic substances are generally fat-soluble but not water-soluble. This is why they are dissolved better at higher alcohol contents, allowing more of them to be carried over into the distillate (i.e., the schnapps) during distillation.

  The following experiment illustrates this effect very clearly: take 3.4 fluid ounces (100 milliliters) each of tasteless alcohol with ABVs of 5 percent, 45 percent, and 96 percent, respectively. Add 0.35 ounces (10 grams) of raspberries to each, and let them sit for about six to eight weeks. After this time has passed, the 5 percent ABV batch will taste a little like raspberries, and the fruits in it will be pink and still relatively flavorful. The 45 percent ABV batch, on the other hand, will have an intense raspberry flavor, while the fruits in it will have become colorless and only vaguely taste like raspberries. The 96 percent ABV batch will taste bitter because the very high alcohol content was able to extract the bitter-tasting elements of the seeds, too.
- The aforementioned "yeast film" (i.e., mold film, a mixture of aerobic bacteria/mold/yeast floating on the surface), unfortunately a common problem with conventional mashes, does not form at all in high-grade mashes. This problem simply doesn't exist.
- A 20 percent ABV mash can't rot or grow moldy thanks to its high alcohol content, so it can be stored unfiltered for several years without adding any preservatives. The mash doesn't need to be used immediately after fermentation is finished. It is in fact a very good idea to leave a high-grade mash standing for at least six months or longer after it's done fermenting because this creates a mild, very clean fruit flavor. This practice also helps prevent possible interfering tastes such as pungency and an off flavor. Doing this with conventional mashes is impossible because without adding preservatives and antioxidants they cannot be stored as long without a drop in quality; a distinctive moldy taste will result.
- Because of the high alcohol content, only one distillation is necessary to reach a high enough alcohol content for schnapps (about 40 to 45 percent ABV) in the distillate. It makes the second stage in two-stage distillation, which diminishes the flavor and requires a lot of effort, unnecessary. Distilling more times

does raise the distillate's alcohol content and make it purer (more tasteless), but it also inevitably causes more of the flavor to be lost in the kettle (for more on this, see chapter 4).

- Last but not least is the yield: if working with turbo yeast, about 30 percent of the mash volume is the amount of distillate (translated to 43 percent ABV) to expect. This is huge in comparison to conventional mashes, which yield about 3–4 percent of mash volume with rowanberries or about 8–10 percent with prune plums. By the way, it is an outright myth that the high amount of alcohol would dilute the taste of the distillate. So with high-grade mashes you will get a higher yield and a more intense fruit taste. Now you can imagine how much aroma is lost when using conventional mashes with low alcohol content and distilling them twice.

Total sugar content of the mash before fermentation in relation to desired alcohol content after fermentation ▾

Desired alcohol content in the mash [% ABV]	Total sugar needed [grams* of sugar/ liter* of mash]
1	23
2	42
3	61
4	81
5	100
6	119
7	139
8	158
9	177
10	196
11	216
12	235
13	254
14	273
15	293
16	312
17	331
18	351
19	370
20	389
21	408
22	428
23	447
24	466
25	485
* 100 gram = 3.5 ounces, 1 liter = 0.26 gallons	

Turbo yeast

Several producers offer a variety of different turbo yeasts, supplied by (home) distiller stores. Unfortunately consumers have reported musty smell with yeasts of a specific brand, as well as a typical "turbo yeast" off-taste in the distillate. So far we haven't observed any problems with the brand Prestige; usually we use "Prestige 8 kg turbo yeast."

Tests have shown (see page 34), that a high-grade mash, stored for at least six months after fermentation, is completely free of heads. So don't be impatient; storing high-grade mashes for at least six months is a must! The longer you wait, the better the results. Contrary to conventional mashes, the schnapps is drinkable directly after distilling and diluting to 43 percent ABV. Distillates of conventional mashes are stored for at least two, better three or four, years before they are bottled and sold. Directly after distillation they are undrinkable because of their strong, pungent taste.

Preparing a high-grade mash

We have already discussed how much water you need, checking the pH level, and adding pectinase. As far as yeast goes, you can use either turbo yeast or sherry yeast. The difference lies in the attainable alcohol contents:

Turbo yeast: at least 20 percent ABV in the mash
Sherry yeast: at most 16 percent ABV in the mash

For example, to get an ABV of 20 percent in the mash, you'll need 13.7 ounces (389 grams) of sugar per quart (liter) of mash. Remember: the sugar is turned into alcohol by the yeast, so no sugar will be left over after fermentation. The table on page 23 gives the amounts of sugar necessary to attain various different alcohol contents.

Example ▸

For an alcohol content of 12 percent ABV, we need 8.29 ounces (235 grams) of sugar per quart (liter) of mash.
For an alcohol content of 19 percent ABV, we need 13.1 ounces (370 grams) of sugar per quart (liter) of mash.

Since the fruit itself contains some sugar (see the section on conventional mashes), this amount can be subtracted from the amount you have to add. If you do not account for this sugar, you may reach an alcohol content well above 20 percent when using turbo yeast. Therefore, it's only sensible to account for the sugar content of the fruit when using sherry yeast.

Example of calculating the amount of sugar ▸

Sherry yeast: Apricots can reach a natural alcohol content of 7–9 percent. We'll use 8 percent in our calculation. This 8 percent ABV corresponds to 5.57 ounces (158 grams) per quart (liter). To reach 16 percent, however, 11.0 ounces (312 grams) per quart (liter) are necessary. We must therefore add 5.43 ounces (154 grams) of sugar per quart (liter).
312 – 158 = 154

The given amount of sugar should not be added to the mash all at once. This would be too much sugar, killing off the yeast through the high osmotic pressure. Apart from that, the more sugar added at the beginning, the heavier the fermentation, and the more the fruit flavors will be blown out by the CO_2. For the same reason is it crucial to watch the ambient temperature, which should not exceed 66°F (19°C). If the fermenting container is larger than about 50 gallons (200 liters), it may even be necessary to cool the container with a cooling unit. Only add a third of the sugar when you're first preparing the mash (don't forget to stir vigorously). Add another third one week afterward at the latest. If the room temperature nevertheless reaches above 68°F (20°C), fermentation will take place much faster and you'll have to add the second third after only a few days or else all the sugar will be used up and the yeast will starve, bringing the fermentation to a halt.

How can you check this? Taste the mash: if it tastes sweet, you can wait, but as we said, the second third shouldn't be added any later than a week after the first third. If, on the other hand, it tastes dry, the next sugar addition must be carried out immediately.

The final third should be added a week after the second third at the latest, or earlier if the mash already tastes dry. It's also true here that forgetting to add the sugar or adding it too late will cause fermentation to cease due to the yeast starving.

Do not make the mistake of adding too little sugar with the reasoning "it's not worth using so much sugar, less would surely suffice." As we already pointed out in the fermentation equation (page 6), every 35 ounces (1 kilogram) of sugar produces about 51 fluid ounces (1.5 liters) of alcohol, translated to 43 percent ABV.

Note: Even if you're adding sugar, never work with unripe fruit. The added sugar does make up for the fruit sugar, but you'll experience a serious loss of flavor compared to if you had use fully ripe fruit.

Two examples to help you better understand how to calculate how much sugar you need:

◂ Example 1

We have 26 gallons (100 liters) of pear pulp and wish to create a 20 percent ABV mash. What ingredients do we need to add?
Solution:

- *The 26 gallons (100 liters) already includes the water used for cleaning the equipment, about 1.5 gallons (5 liters).*
- *Pectinase: 6 milliliters for 26 gallons (100 liters) of mash.*
- *Turbo yeast: 4 ounces (115 grams) for 26 gallons (100 liters) of mash.*
- *Sugar calculation: Our desired alcohol content is at least 20 percent ABV, so we'll need 13.72 ounces (389 grams) of sugar per quart (liter) of mash, see table on page 23. The sugar content of the fruit itself is intentionally ignored to reach an ABV over 20 percent.*
- *We add the first third of sugar, 28.7 pounds (13 kilograms), immediately.*
- *pH value: Add 10 fluid ounces (300 milliliters) of Biogen M to adjust the pH to 3.5, then check it with a pH strip and add more Biogen M if necessary.*
- *Stir everything vigorously and seal the fermentation container with a fermentation lock.*
- *Further steps for the mash: After a week, add the second third of sugar, 28.7 pounds (13 kilograms) again. After another week, add the final third, meaning that we've added a total of 86 pounds (39 kilograms) of sugar.*

Example 2 ▸ *We want to ferment 26.4 quarts (25 liters) of the mash to an ABV of 16 percent using sherry yeast. How much sugar will we need?*
Solution:
For an ABV of 16 percent, we need a total of 11.0 ounces (312 grams) of sugar per quart (liter) of mash, see table on page 23. Pears already contain enough sugar to attain an ABV of 5 percent, which corresponds to 3.53 ounces (100 grams) of sugar per quart (liter).
$(11.0 - 3.53) \cdot 26.4 = 197.21$ ounces of sugar (12.3 pounds)
$(312 - 100) \cdot 25 = 5{,}300$ grams of sugar (5.3 kilograms)
We must therefore add a total of 12.3 pounds (5.3 kilograms) of sugar, one-third of it, or 4.1 pounds (1.8 kilograms), immediately.

The fact that we're working under acidic conditions (pH = 3.0–3.5) has a second benefit for high-grade mashes besides suppressing faulty fermentation: cane or beet sugar (i.e., common refined or table sugar) belongs to the disaccharide group and is called sucrose in technical terms. Disaccharides are composed of two sugar rings—hence the "di-" prefix—and cannot be fermented directly. Under acidic conditions, disaccharides are broken down into two single rings (i.e., monosaccharides), a group that includes, among others, glucose and fructose (which we know from the fermentation equation). These monosaccharides are easily fermented by yeast.

Since a package of turbo yeast also contains nutrients, you don't necessarily have to ferment a nutrient-rich fruit pulp—you can also use it to produce pure, tasteless alcohol. To do this, ferment a mixture of water and sugar (see page 98 for the recipe) and distill the result. This kind of alcohol can be used for infused alcohol or spirits, among other things (see chapters 6 and 7).

Acid can break disaccharides down into monosaccharides ▸

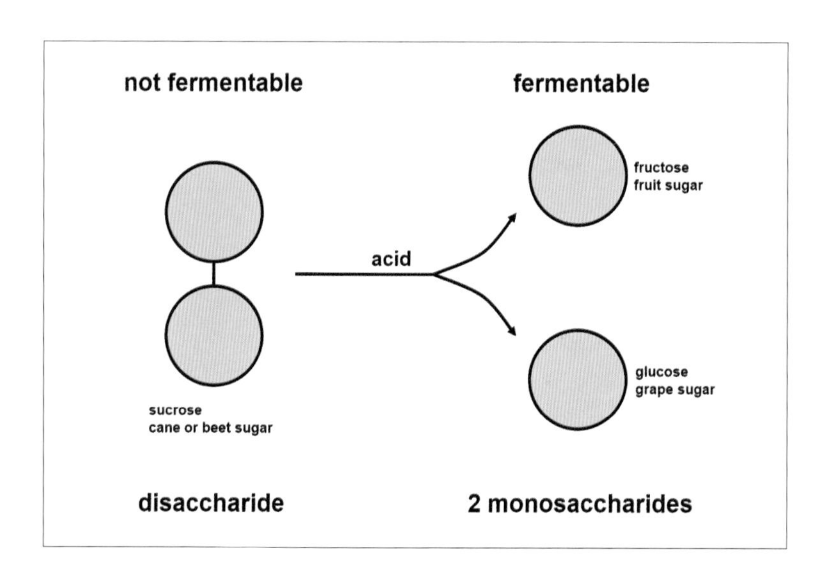

The following table summarizes the process of making a high-grade mash once again. For other amounts, adjust the dosages accordingly. Since turbo yeast is a mixture of several different substances, you should always use at least a teaspoon of it in total.

Preparing a high-grade mash

Fruit pulp	Clean and wash the fruit, remove stems and rotten areas, and mash it. With drupes, make sure that no pits are destroyed.
Water	Use as much water as is necessary to clean the equipment (at most a third of the fruit pulp). Exceptions are noted in the recipes.
Pectinase	Verflüssiger Spezial: 0.1–0.8* fluid ounces per 25 gallons (3–25* milliliters per 100 liters) of mash
pH correction	Measure the pH value, correct it to 3.0–3.5 if necessary, e.g., with fluid acid blend Biogen M: 10 fluid ounces per 25 gallons (300 milliliters per 100 liters) of mash
Yeast	Turbo yeast: 4 ounces (115 grams) per 21–26 gallons (80–100 liters) of mash. No additional nutrients are necessary. *or* Sherry yeast + yeast nutrients: follow the manufacturer's given dosage
Added sugar	A third of the calculated amount immediately Second third about a week later Final third another week later
Fermentation	Let it ferment with a fermentation lock. Optimal temperature: 61–66°F (16–19°C). Stir once or twice a week, stop stirring when the fruitcake has sunk, then the fermentation is finished (after about 2–3 months).
Storage	Store fully fermented mash at least about 6 months for a heads-free distillate with rich flavor and smooth taste. No need to store the distillate (see section storage).
* The amount depends on the fruit. Soft fruit like drupes need less pectinase than hard fruit like quinces, rose hips, or rowanberries. Roots like Jerusalem artichokes need 0.8 fluid ounces per 25 gallons (25 milliliters per 100 liters).	

▲ 20 percent ABV with turbo yeast, 16 percent ABV with sherry yeast

Checking the progress of the fermentation

What should you check, and how often?

This section applies to both conventional and high-grade mashes. An ideal mash fermentation requires that you check its progress very exactly in order to be able to take measures against any problems that might arise before it's too late.

The very first thing you should do is make sure that you don't fill the fermentation container more than three-quarters full. Otherwise, it will overflow during fermentation. We're speaking from experience here. A prune plum mash of ours was undergoing furious fermentation (the first two to three days) excellently, but the barrel was more than three-quarters full because we decided to add the last of the harvested fruit as well. We noticed during one of our routine checks that some liquid had escaped through the fermentation lock. We prudently removed the barrel from the basement out into the open, and the full extent of the fiasco became clear to us when we tried to open the fermentation lock. Just briefly touching the lid was enough for it to fly up into our faces, along with the prune plum pulp. It was impossible to stop the fountain erupting from the barrel; it sprayed and sputtered. An enormously excessive pressure had developed in the barrel because a prune plum pit had stopped up the fermentation lock. Conclusion: we lost a lot more of the prune plum mash than we would have by simply not adding those last fruits into the barrel.

> ☺ Tip: Always stir in the yeast after all the other ingredients, particularly after checking the pH. If you adjust the pH of a mash that's already fermenting by adding acid, the shock can cause the fermentation to come to a standstill.

> ☺ Tip: If you're using (liquid) sherry yeast, you may encounter difficulties getting fermentation to start. In this case, therefore, adding a yeast starter is a good idea (see page 35, "problems during fermentation").

Daily check

While the fermentation is taking place, you should check the fermentation container every day and note whether the fermentation lock is still bubbling (as previously mentioned, the bubbling frequency will decrease gradually). This is the best way to determine whether the fermentation is proceeding properly. You should also check the room temperature, as an excessively high or low temperature can very quickly stop fermentation. It should never drop below 59°F (15°C) and, for maximum flavor, should never rise above 68°F (20°C). The yeast die off above 82°F (28°C). Be careful if you leave the barrel outside because nighttime temperatures can fall below 59°F (15°C) even in the summer.

You should also perform a taste test. A conventional mash, one without added sugar, should be tasted to check against mold and faulty fermentation. You will perceive their presence immediately; it tastes like vinegar or rotten fruit. If you're making a high-grade mash—one with added sugar—you must also take note of the mash's sweetness. If the mash tastes sweet, there's still enough food available for the yeast. If it tastes harsh and dry, on the other hand, you must add sugar immediately, as long as you haven't already added the entire calculated amount of sugar. Otherwise the fermentation will quickly come to a standstill due to the yeast starving.

Regular stirring

As long as it is fermenting, it's a good idea to vigorously stir the mash once a week (smaller containers can be shaken vigorously). Strongly push the fruitcake that forms at the top down to the bottom and mix it well with the liquid. This is the only way to keep the solid components that contain the flavor in contact with the liquid and allow the flavors to transfer. If you don't stir the fruitcake, it will dry out over time and give the mash a straw-like taste. Fermentation won't have concluded until the fruitcake, which will have been largely broken down by the yeast in the meantime, has sunk to the bottom. From above a clear liquid is visible, sometimes—depending on the fruit—only a few centimeters deep. At this point you don't need to stir any more.

▲ Homemade stirring plate

Mashes of pomaceous fruit especially, but often also other fruit mashes, still contain small parts of fruit right after mashing. These parts get soft during fermentation, but they do not dissolve completely. For the last few years we have been mixing every mash intensely a week after fermentation start with a mixer attachment for power drills. After this treatment, the mash is literally a cloudy fruit juice with a few pips floating in the juice, the perfect mash! Recently we found out that professional producers of high-quality fruit brandy do the same.

☺ Tip: You can build the best stirring tool yourself: take a broomstick and bolt it to the middle of a wooden board. Push this "stirring plate" up and down in the fermentation container; this will do an excellent job of mixing up the mash.

Mash production in pictures

▲ This is what you need to make a mash: fruit, turbo yeast, liquid pectinase, Biogen M, pH strips, sugar, a mash container, a fermentation lock.

▲ The fruit is being washed

▲ and mashed.

▲ Add very little water in order to maximize the flavor.

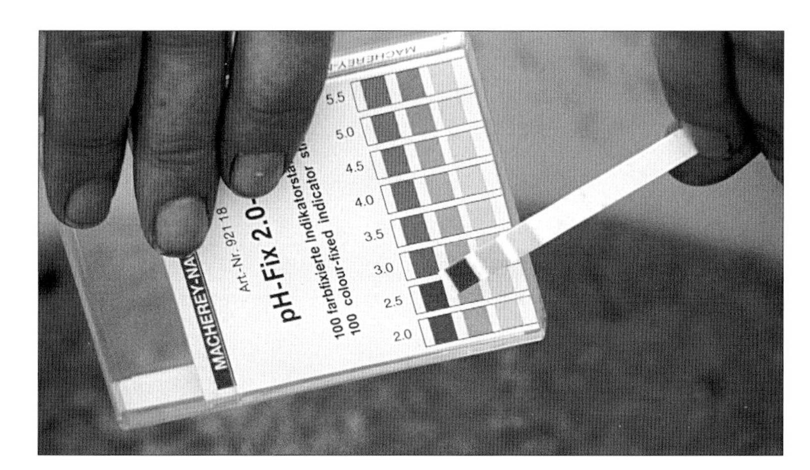

◀ pH check and addition of Biogen M if necessary.

▲ Add pectinase, yeast, and sugar.

▲ Stir everything thoroughly,

▲ attach the fermentation lock,

▲ and don't forget to make a record.

Weekly measurements

You should measure the alcohol content of the mash every week with a vinometer. This allows you to tell whether your mash is making progress. If you're using citric acid, you should also check the pH value every week so that you can take note of changes promptly and add more acid if necessary.

Checking the mash ▸

Daily	Is it still bubbling? Is the room's temperature okay? >59°F (15°C) and <68°F (20°C)
Every 3–7 days	Stir (as long as it's still fermenting) Taste: No bad tastes; check the sweetness in high-grade mashes
Weekly	Alcohol content with vinometer pH (only with citric acid)

Measuring the alcohol content of the mash with a vinometer

A vinometer is essentially a narrow glass capillary tube. The liquid will sink to a different level in it depending on the alcohol content. If you want to take a measurement from a mash, first filter about 1–2 tablespoons of it with a coffee filter, as solid particles can clog the tube. Fill the vinometer's funnel completely with the filtrate. Now the tube will slowly fill with liquid. Wait until the liquid is dripping out the bottom and check whether the tube is also free of small bubbles. It's best to hold the vinometer up to a light so that you can see the bubbles more easily. If there are any small bubbles, gently suck the bottom until the tube is completely full of liquid. Then quickly turn the vinometer 180° vertically so that the funnel is at the bottom and the liquid drips out of the funnel. The liquid in the tube will sink downward, and the height of the fluid level will correspond to the alcohol content. About five seconds after you turn it over, you'll be able to read the alcohol content directly in ABV. Be sure to thoroughly rinse out the vinometer with distilled water after you use it or it may become clogged by leftover material that dries out.

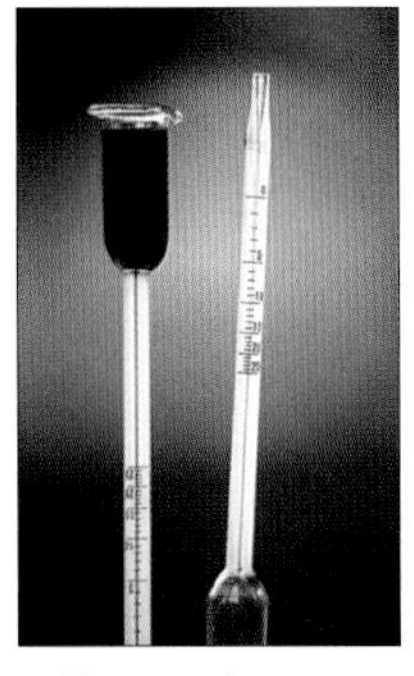

▲ How to take measurements with the vinometer: fill it (left) and take the reading (right)

Dissolved substances in the mash, like sugar, cause a vinometer to give somewhat inaccurate readings. We carried out a series of measurements in which we compared pure alcohol to alcohol with dissolved sugar. The inaccuracy comes to approximately +/- 2 percent ABV, which is close enough for this method to be useful for our purposes. We've observed that the displayed value is generally too high in concentrated sugar solutions, so if the mash tastes sweet, the true value is approximately 2 percent ABV lower than the displayed one.

Name		Launch date		O finished
		Fermentation start		
		Fermentation end		
		Storage time		

General data

Description		
Fruit pulp		Total volume
Water		
Yeast		
Pectinase		
Yeast nutrients		
pH-Control		
Temperature		
Sugar – total to add		
Sugar – first portion		

Fermentation progress

Date	Alcohol content	Temper-ature	Sugar		Actions, taste, appearance, notes
			added	total	

◂ Fermentation log

Fermentation logs

It's generally a good idea to maintain a precise log of fermentation. For one thing, it makes it easier to recognize potential issues that may arise later. If, on the other hand, your mash goes especially well, you can follow the process you used exactly in the next year. If you're making many different types of mashes, keeping a log or notes is essential to avoid losing track of what you've already done to the various different mashes. The adjacent sample log gives you an idea of how yours might look.

When is the fermentation complete?

The following indicators show that fermentation has ceased. However, they cannot tell you whether fermentation is actually complete or has just been interrupted. All you can say is that it is no longer fermenting.

- The fermentation lock is no longer bubbling. (Note: toward the end of fermentation, it will only bubble very occasionally. The

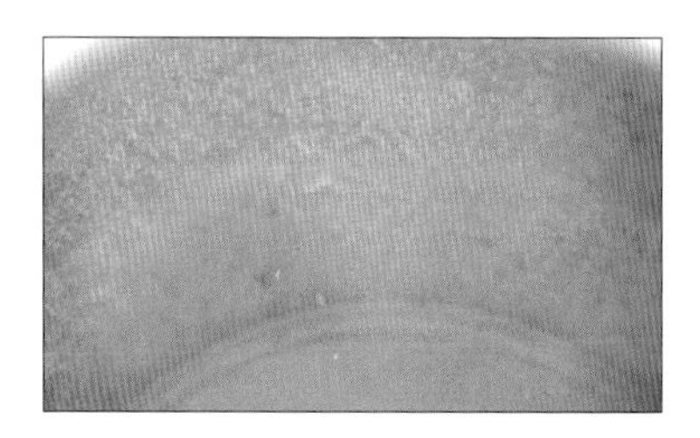
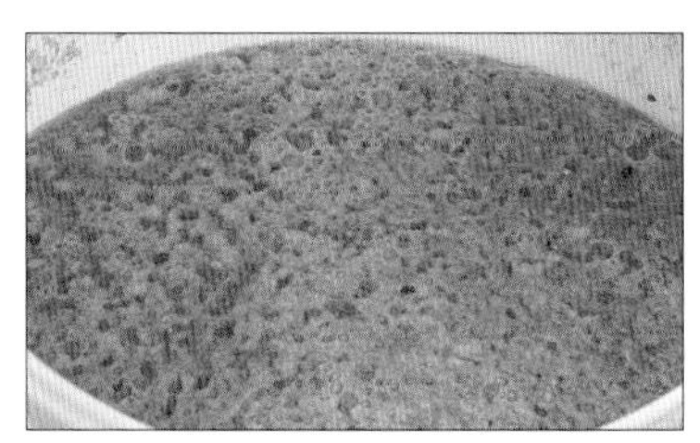

Finished (left) and still fermenting (right) mashes ▸

barrel may also contain leaks, which will also cause it to stop bubbling.)

- The mash no longer foams like champagne when it's stirred.

If you observe both of the indicators given above in addition to the following criteria, fermentation has completed successfully.

- The vinometer indicates the desired alcohol content.
- The fruitcake is no longer floating on the surface and all solid components have sunk to the bottom. A clear, wine-like liquid is present above them. The fruit flesh has also largely disintegrated, especially if you used pectinase, leaving only a small amount of solid material.
- The mash tastes dry (in high-grade mashes, the taste can often still be somewhat sweet after a successful fermentation, that's normal).

Once the mash has been successfully fermented, the most important and most difficult step in producing an excellent schnapps is complete. Since a good mash is the only thing that can guarantee good schnapps, there's (almost) nothing that can go wrong during distillation at this point. If, however, the mash tastes tainted or like mold, vinegar, or anything similar, it will not be possible to make good brandy from it. If the mash tastes so good that it almost feels like a waste to use it in distilling, it's perfect!

The optimal time for distilling

After fermentation is complete, the mash should be left standing without any stirring for at most one to two months before the distillation. This is when the esterification processes (reactions of alcohol with organic acids) take place in the mash, which leads to a more intense flavor in the distillate. It's actually generally true that the longer you a store a mash, the better the resulting schnapps will be. Among other effects, doing so helps you avoid unwanted flavors like pungency or a "yeasty" off smell. However, conventional mashes (with less than 16 percent ABV) cannot be stored for any longer than the one to two months mentioned earlier without preservatives and antioxidants or else the resulting schnapps will have a "moldy basement" flavor. Since high-grade mashes do not have this issue, they should be left standing for at least six months before the distillation. Storing the mash for a full year makes the schnapps incomparably mild and flavorful, partly because a major component of the heads, acetaldehyde (an intermediate product of alcoholic fermentation), breaks down completely during storage. This means that clean, high-grade mashes are usually completely free of heads after being stored long enough.

Problems during fermentation

Sometimes fermentation ceases. This is always due to one of the following causes:

- Temperature is too high: >82.4°F (28°C)
- Temperature is too low: <59°F (15°C)
- Too much sugar
- Not enough sugar
- Too acidic: pH < 2.5
- Old yeast

If you keep these potential issues in mind, you should hardly ever encounter the issue of fermentation stopping. If it does still happen, fermentation can be restarted as follows. First of all, you must determine whether any alcohol has already formed in the mash or if fermentation never started at all and there's no alcohol in it. The best way to check this is with a vinometer. If you're still not sure whether alcohol has formed or not, it's best to just carry out the second process.

No alcohol has formed in the mash yet:

If fermentation hasn't started twenty-four to forty-eight hours after you've added the yeast, the yeast was probably too old, so all you need to do is add new yeast to the fermentation container and stir. Always check the expiration date when you're buying yeast. Another frequent cause is low environmental temperature. To prevent this, start fermentation at about 72°F–75°F (22°C–24°C) ambient temperature. As soon as the mash is fermenting furiously, carry the container with the mash into a room of 59°F–66°F (15°C–19°C). Make very sure, however, that you have solved whatever issue came up in your previous attempt. If you don't, it won't work this time either. Yeast proliferates with air, so sometimes, if the yeast wasn't too old or the temperature wasn't too low, it helps to mix air into the mash twice per day with a hand blender (only this kind of mixer works) or with a mixing attachment for power drills for two or three days. Stop this treatment when fermentation is clearly recognizable. If after this period the mash is still not fermenting, it doesn't make sense to go on; the probability is rapidly growing that wild yeasts, with all the negative consequences they bring, have taken possession of the mash.

Alcohol has already formed in the mash:

In this case, adding more yeast won't do any good because the yeast (directly from the vial or the packet) will die off as soon as they come into contact with alcohol, even if the alcohol content is still very low. This is true of both liquid and dry yeast. They can only survive an alcoholic medium once they've adjusted to its presence. You'll thus have to add a yeast starter, for example one of the two following recipes.

If you're working with a high-grade mash, do not forget to add the corresponding amount of sugar (if this has not already been done). If the mash already has an ABV above about 14–16 percent, you will not be able to restart fermentation even with a yeast starter.

Yeast starter recipe I:
This is nothing but a half-empty 1-quart bottle (1-liter bottle) with pasteurized, preservative-free apple juice and the desired yeast (1/8 of a sachet of turbo yeast is sufficient, any other yeast according to the manufacturer's instructions). Close the bottle with a fermentation lock and set it aside in a warm place (approximately 72°F or 22°C). Because yeast proliferates with air, remove the air lock twice a day, close the bottle tight, and shake it. Don't forget to reattach the air lock. After a few days, when a furious fermentation is recognizable, mix the contents in the mash, half a bottle per 5 gallons (20 liters) of mash.

Yeast starter recipe II:

- The volume should be about 2–5 percent of the main batch, the listed amounts of the following additives are calculated for 2 quarts (2 liters) of lukewarm tap water
- 10.6 ounces (300 grams) of sugar
- 1/4 of a packet of turbo yeast or any other yeast according to the manufacturer's instructions
- Adjust the pH (the same way as with the "big" mash) or add the juice of half an orange and half a lemon, as long as its taste won't interfere undesirably.

Pour everything together and stir vigorously. After one or two days, depending on the temperature, furious fermentation will begin. Add the yeast starter in this condition—when everything is foaming and fizzing, still during the furious phase—to the mash that is no longer fermenting (don't forget to stir) and the problem should be solved.

Note: As a general rule of thumb, you should use approximately one-twentieth the quantity of your original mash for your fermentation starter. If you have a 10 gallons (40-liter) mash that's no longer fermenting, you'll need about 2 quarts (2 liters) of fermentation starter.

Filtering the mash

Some distillers filter their mashes, which is a big mistake—unless they're making a fruit wine (see chapter 5). The skins of the fruits contain most of the flavor. If you filter them out, the schnapps will taste much more "empty" than an unfiltered batch. It is possible for solid components to burn, but chapter 3 describes some very easy methods of avoiding this problem.

As a general rule, therefore, you should not filter the mash. If you want to filter it anyway, you'll find some tips and tricks in chapter 5 (in the section on filtering mashes).

Storing the mash

High-grade mashes (i.e., those with an ABV over 16 percent) can be stored under cool and airtight conditions for several years without any issues. As we've already described, this is a very good idea in terms of quality, so high-grade mashes should be stored for at least about six months before distillation. It's best to use a plastic container as they allow in less oxygen than wooden ones.

Conventional mashes should be used at the latest after one to two months as they run a high risk of rot, mold, and vinegar formation. If you want to store them for longer instead, you'll need to use a fixing agent. Adding glucose oxidase, 0.07 ounces per 25 gallons (2 grams per 100 liters) of mash, and cellulase, again 0.07 ounces per 25 gallons (2 grams per 100 liters) of mash, can minimize the amount of oxygen that makes it into the container, which is the cause of undesirable changes. Be certain to use a plastic container and fill it as full as you can to leave as little room for air, and thus oxygen, as possible. Store the container under cool conditions.

Making a mash from grains, corn, or potatoes (starchy products)

Making a mash from starchy products is drastically different from making a fruit mash because fruits, as opposed to grains or potatoes, contain the monosaccharide fructose (fruit sugar), which can be converted into alcohol with the help of yeast. Starchy products do not contain monosaccharides or disaccharides, but rather the polysaccharide starch (see picture on page 38). Starch is a molecule that is made up of a large number (between 100 and 14,000) of chained-together glucose molecules. To get the monosaccharide glucose (grape sugar), the chains must be broken down. Only at that point is fermentation with yeast possible.

- Grain mash with amylase: The starchy product is first finely ground (grains) or finely sliced (potatoes). You can of course

Note: Grain mashes can also be fermented as high-grade mashes, of course. This has the advantage of bringing out more of the grain flavor and only necessitating a single distillation. To do so, follow the instructions in the section on high-grade mashes with added sugar.

also use flour of any type. To convert the starch into fermentable sugar, the mash must be heated. Starchy products contain very little water, so you should mix them with water at a 1:1 ratio. First heat the water to about 195°F (90°C) and then add the grain or potatoes—the temperature must be at least 158°F (70°C) when you do so. Then let the mash cool down to about 140°F (60°C), adjust the pH to 5, and add the enzyme amylase to it (the amylase breaks the starch down into the fermentable sugar). Leave the mixture at 140°F (60°C) for about two hours and stir it occasionally. Follow the dosing instructions on the packaging precisely.

Once it's cooled down below 81°F (27°C), add pectinase and yeast (at the same dosages given for fruit mashes). During fermentation and distillation, just treat the mash like a fruit mash.

- Grain mash from homemade malt: Put the grains on a baking sheet and cover them with a wet cloth or spray them with water and keep them moist until they germinate (about three to five days). Then dry them in the oven at about 122°F (50°C) for at least two hours. After that, thoroughly grind the malt and mix with water, using 3–4 liters of water per kilogram of malt. Heat the mash up to 95°F (35°C) and mix it intensely with a blender. Heat it further to 140°F (60°C), but be careful not to exceed 149°F (65°C), and leave it there for at least fifty minutes. Then heat it further to 158°F (70°C), but not past 167°F (75°C), and leave it there for an additional fifty minutes at least. Next, let the mash cool down to 77°F (25°C) and adjust the pH to 3 (see the section on fruit mashes). Aerate it by stirring vigorously or pouring into another container. Add yeast and pectinase, about 0.8 fluid ounces per 25 gallons (25 milliliters per 100 liters), as described in the section on fruit mashes.

Starch decomposition ▸

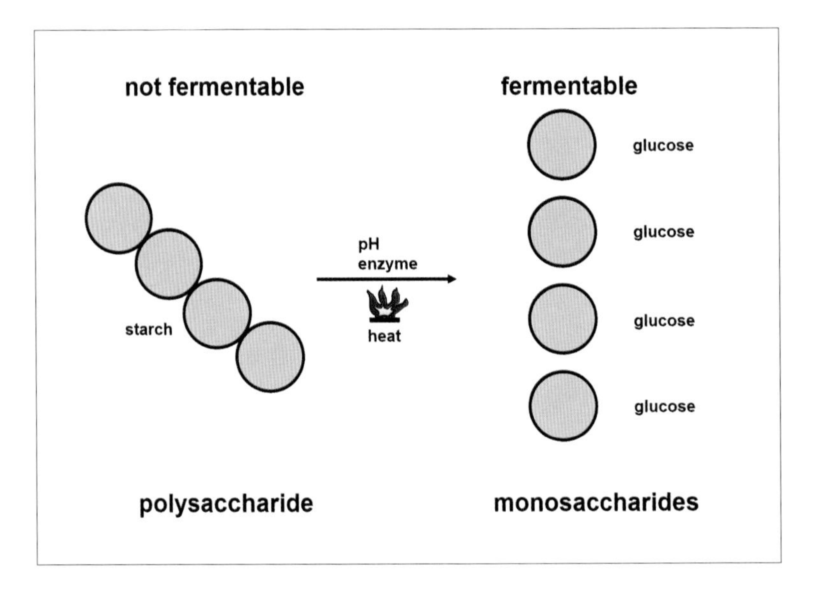

Preparing the mash	• Grind grains, cut up potatoes • Add an amount of water equal to the amount you're fermenting • Heat the water to 195°F (90°C) • Stir in the grains/potatoes • Let it cool down to 140°F (60°C) • Add amylase • Let it cool down below 81°F (27°C)
pH	• Adjust to 5, while the mash is at about 140°F (60°C)
Pectinase	• Add three to four times as much as indicated, unless you're using a special pectinase for starchy products
Yeast	• Any kind of described yeast

◂ Mash from starchy products

☺ Tip: Leave potatoes outside in the frost or put them in the freezer. This partially breaks down the starch and makes the taste sweet. It's somewhat possible to ferment them afterward, but with a limited yield.

CHAPTER 3

Stills

In this chapter, you will learn how to construct a distillation apparatus as well as what you need to look out for if you purchase a still, as not everything sold as a "still" actually functions as one. We will also explain how a still works and the proper way to use one.

Principles of construction

Every still, regardless of how complex, works by means of the same principle: a mixture of alcohol and water (mash, wine, crude distillate, etc.) is brought to a boil. This produces steam, which is itself also an alcohol-water mixture, but with a higher alcohol content than the liquid due to a physical law of nature that you can learn more about in chapter 4. If this steam cools, it will again become a liquid (i.e., it will condense). But this time, the liquid—also called condensate, distillate, brandy, or schnapps—will have a much higher alcohol content than before. The pot containing the boiling liquid is known as the "kettle," the pipe that leads the steam upwards is the "column," the pipe that leads steam downwards is the "lyne arm," and the device for cooling the steam is the "condenser" or the "cooler." The trick is to carry as much of the fruit flavor as possible through the steam and into the distillate.

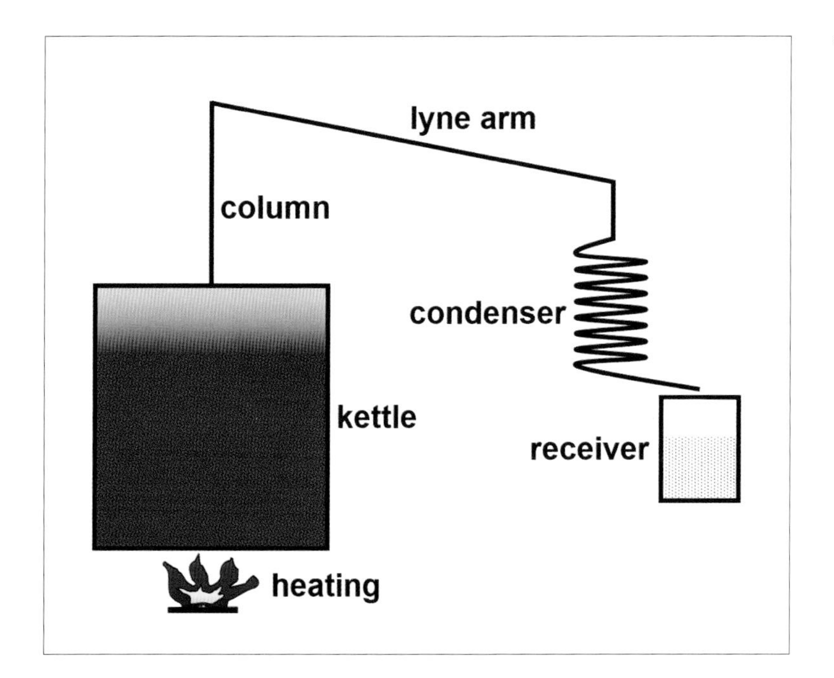

◂ Basic still setup

Some key points

If you're considering purchasing a still, there are some very important points you should keep in mind. The price alone tells you nothing about whether the still will actually function. There are very expensive "professional" stills that are unfortunately completely wrongly made. To spare yourself future frustration, you should be absolutely certain to pay attention to the following factors when you make your purchase.

1. The apparatus must include a steam thermometer. If it doesn't, it will only have limited functionality as you will not be able to separate heads and tails exactly.
2. The thermometer must be affixed to the highest point where uncooled steam will be. This is the only location where the steam's temperature corresponds to the alcohol content of the distillate, and this is the only way to compare values from differently constructed stills. If the thermometer is attached lower down, you may encounter deviations of twenty degrees Fahrenheit (ten degrees Celsius) or more.
3. The width of the steam pipe should be as great as possible. The narrower and longer the pipe, the higher the pressure drop of the still. The consequence of this is that there will be a buildup of excess pressure in the kettle, which will drastically alter the steam temperature, destroying the sensitive flavors and aromas.
4. The materials used in the still must not affect the taste of the steam or the liquid. Heat-resistant glass, stainless steel, or copper, for example, are particularly suitable. You should avoid rubber seals that are not chemical resistant because they will give the distillate a rubbery taste. Highly reliable sealing materials include cork, teflon, and silicone.
5. The still should include a steamer basket made of stainless steel or a jacketed kettle to prevent the content from burning. A steamer basket is the cheaper option, but it is still very effective.
6. To prevent considerable losses in aroma and yield, the empty space between the boiling surface of the kettle content and the still-head should be as small as possible. Therefore, the kettle should be filled up to at least three-quarters of its height.

Materials

The right material for your purposes depends on various criteria. In any case it has to be hard enough to be stable, heat-resistant, resistant to fluid alcohol as well as alcohol steam, resistant to weak acids, and it should be food-proof. Without any limitation, there are in fact only three materials possible: copper, steel, and heat-resistant glass (silex). Due to the weak acids, aluminum is suboptimal. If it is used, it must be only for the kettle, not for the still-head or the other parts. Of course, it's also possible to choose different materials for different parts of the distilling device, for example, using a copper kettle and a steel condenser or a steel kettle and a glass condenser.

If you are planning to construct a still by yourself, the most suitable material depends on your mechanical skills. Welding stainless steel is unfortunately not so easy. If you wish to use copper, use brazing alloy (silver solder). Verdigris mainly forms on soft copper; hard, high-density copper generally doesn't have this problem. For small stills made of glass, you can find all necessary parts at laboratory suppliers. The parts have standardized gas-tight connections, so they can easily be stuck together. With a little innovative spirit and a few mechanical skills, it's not difficult to construct a condenser suitable for pressure cookers and therefore replace the pressure valve with the column or lyne arm inlet. You'll find hundreds of pictures of different self-constructed stills on our webpage (see web address at end of the book).

Many sources will tell you that the ideal distilling kettle or the entire still should be made of copper, or that at least the still-head should be. Copper was traditionally used because it can be easily processed and is able to degrade cyanides. These substances develop, as previously mentioned, from broken pits.

Tradition and easy processing: every common village smith of yesteryear could process copper, but not stainless steel. So stills were made out of copper for centuries. Today, high-tech stills on an industrial scale are made of stainless steel. Only Scotland is different: if a distillate wasn't distilled with a copper still, it's not allowed to be called Scotch whisky. Some people have reported a difference in taste between Scotch whisky distilled with a steel still in comparison to a copper still, but this hasn't yet been verified by further studies.

The simplest way to avoid significant concentrations of cyanides is to avoid destroying the pits of drupes. If this isn't possible due to large quantities of fruits and industrial scale of production, cyanides can be degraded using several methods. In any case, it is not sufficient to simply use a copper still, even if the inner surface was cleaned thoroughly. If you picture a copper still, you'll realize that only a very small portion of the steam comes into contact with the sides (i.e., the copper). The rest of it flows through the inside, without touching the sides, until it makes it to the condenser. Professional stills therefore are equipped with a "copper catalytic converter" (although copper doesn't act as catalyst in this matter). This is nothing but a container, approximately the same height as width, packed with copper meshwork, plates, shavings, or something similar, connected between the steam outlet of the still-head or column and the condenser inlet. Flow direction within the container is from the bottom to the top. According to the manufacturer, this converter improves the taste of the distillate for all kinds of mashes, not only with drupes or when containing cyanides. Another method for degrading cyanides is adding copper salt (Cyanurex) to the mash in the kettle, with the disadvantage that the mash is considered hazardous waste after distilling due to the high concentration of copper. An alternative is a kind of heated gas washing bottle, where the steam is passed through a solution of copper(I) chloride before it reaches the condenser. For home distillers the most convenient way to degrade cyanides is to fill the still-head and the column with copper wool, but, as we mentioned, this is only necessary if using mashes containing many broken pits.

"Normal" (red) rubber seals are totally unsuitable for the seals and the plug for the thermometer. The aggressive alcoholic steam, which attacks the thermometer plug especially hard, releases a variety of chemicals from rubber, which would allow you to produce a "wonderful" rubber schnapps. You should therefore only use a silicone or cork plug for the thermometer. Chemical-resistant sealing rings (usually black or gray) are best for the seals because they do not come into direct contact with the alcoholic steam. Alternatives are twisted Teflon tape (PTFE thread sealant tape), a lengthwise-cut silicone tube, or a mixture of flour and water. This mixture hardens when heated, so simply smear it over the leaky area.

Kettle

You don't necessarily need to buy an expensive jacketed kettle (see next section). Its only purpose is to keep the mash from burning, and this can be sufficiently accomplished by placing a sturdy metal sieve or mesh over the bottom of the kettle. It should be mounted on a metal framework that fits the shape of the kettle, at a height of 1–2 inches (3–5 centimeters), and the mesh openings should be approximately a couple of millimeters. If it is placed into the kettle and the kettle is then filled with mash, the mash will not burn. To improve its functionality, the insert can also be covered with a metal fly screen (see picture). Some stills are also fitted with an agitator, but these varieties are relatively expensive. The steamer basket described above works at least as well.

Kettle with a burn protector and an outlet nozzle ▾

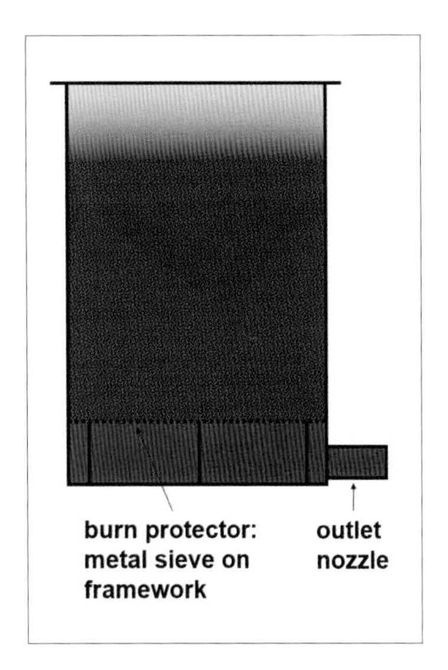

In stills with a kettle volume of more than 2.5 gallons (10 liters), the kettle is usually fitted with handles to make it easier to empty out the contents after distilling. Kettles with a volume of 10 gallons (40 liters) or more generally include an outlet nozzle just above the bottom of the kettle. It will have a diameter of at least 2–4 inches (5–10 centimeters), depending on the size of the still, to keep it from clogging.

▲ Burn protector: a metal sieve covered with a metal fly screen

Heating and jacketed kettles

Depending on the price, size, and layout of the still, the kettle is heated either with an open (wood) fire, a spirit lamp or gas burner, electrically, or through a jacketed system. In jacketed kettles, the jacket itself can also be heated in various different ways. In principle, it doesn't matter how you bring the mash to a boil and keep it there as long as the following criteria are met:

- The maximum heat output should be suitable for the volume of the kettle
- The mash must not burn
- The heat level should be infinitely variable (even at low heat outputs) and above all it should be able to adjust the released heat quickly

If you want to make high-quality schnapps, it is crucial to not boil the mash too intensely so the kettle content will simmer only. The distillate, depending on the volume of the kettle, should only flow out of the condenser at a moderate pace. In small stills with a kettle volume of up to 2.5 gallons (10 liters), it should flow out at a fast drip, not a rivulet yet.

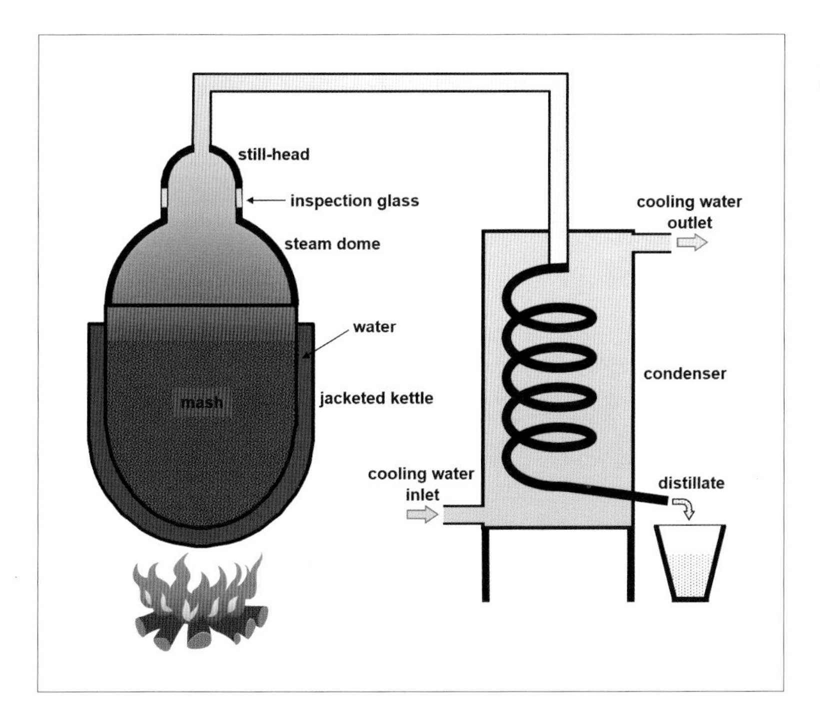

◀ Still with a jacketed kettle

To make sure that you can set the drop frequency (i.e., the distillate flow) as quickly and precisely as possible (infinitely variable), the heat source must not be sluggish like an electric stove, for example, where it takes a few minutes for a change in the heat setting to take effect.

A jacketed kettle has a second kettle surrounding the "normal" (inner) kettle. The space in between is filled with water or oil, which keeps the mash in the inner kettle from burning when it's heated. This is the only reason why a jacketed kettle is used.

In some stills, the jacketed kettle is open. In these cases, using water doesn't really work because it will only be able to reach a maximum temperature of 212°F (100°C). This is not high enough to carry out the full process of alcohol distillation and reach the beginning of the tails (see page 63). For the same reason, it's not a good idea to place a simple kettle in a water bath (as a sort of improvised jacketed kettle). In a closed jacketed kettle, on the other hand, there's a buildup in pressure when heated, like with a pressure cooker, allowing the temperature between layers to reach about 248°F (120°C).

Drawbacks: For one, very high initial costs, and for another, the system is very sluggish. Changes in the heat output only take effect after a considerable delay, like with an electric stove but to a much more extreme degree. Because of this, professional stills that include a jacketed kettle are fitted with cold water nozzles, allowing the temperature between the kettles to change within seconds.

In order to protect the aromas as thoroughly as possible when distilling, it's best to use an infinitely variable spirit lamp or gas burner as a heat source in combination with a burn protector or closed jacketed kettle with cold water injection nozzles.

Column, lyne arm, and measuring the temperature

Reflux = distilling multiple times

There are a variety of different types of columns and lyne arms in existence, depending on how the condenser is constructed. In most cases, the condenser is to the side of the kettle. Depending on the kettle, the pipe is either inclined gently downwards to the condenser right after it leaves the lid (variant 1) or it first goes directly upward, the so-called column (variant 2), before turning in a downward direction (see picture on page 47).

In order to lose as little flavor as possible during the distillation, the column in variant 2 is often fairly short. A vertical pipe performs the function of a reflux or rectification column (i.e., repeated distillation), which has undesirable effects on the flavor. A reflux condenser does, however, more effectively separate the alcohol from the other components, resulting in a cleaner distillate with a higher alcohol content, much like if you were to distill it multiple times consecutively (as in "double-distilled" brandy), but this more

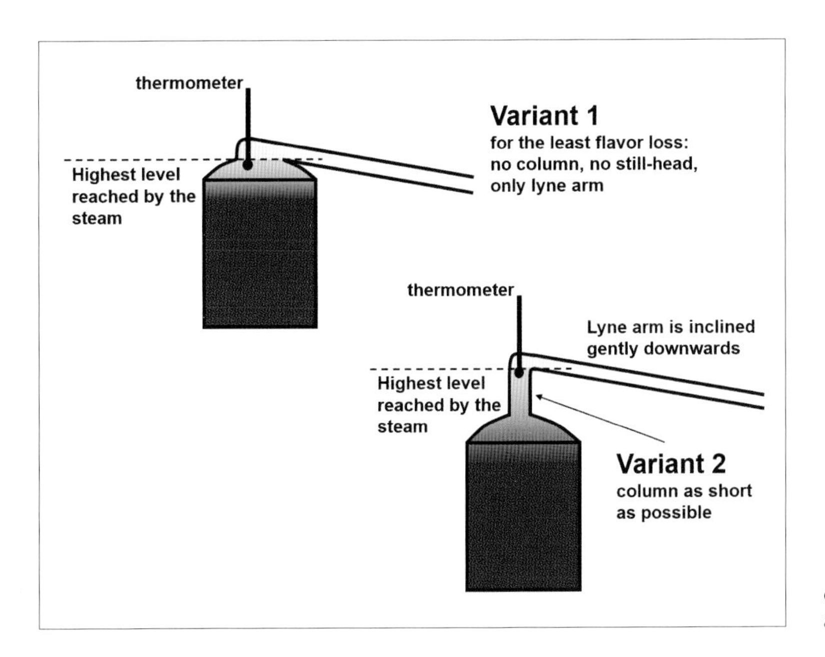

◂ Different ways to construct the column and lyne arm

thorough separation always leaves some of the flavor behind in the kettle. Thus, the cleaner a distillate, the less taste it will have. Many stills are fitted with a still-head (see above) and/or a lyne arm that rises at an angle. The drawback of this setup is that it has a rectifying effect, leading to flavor loss.

The diameter of the column and the lyne arm must not be too small, especially right after the kettle lid. For stills of up to 5 gallons (20 liters), the diameter should be at least 0.8–1.2 inches (2–3 centimeters). The pipe can also enter the condenser conically. Many stills are fitted with a lyne arm that is too narrow. The smaller the diameter and the longer the pipe, the higher the still-specific pressure loss (i.e., the more resistance the steam must counteract in order to reach the condenser). A high pressure loss causes the pressure in the kettle to increase while boiling, and as in a pressure cooker, the boiling and steam temperature will also increase. The higher the temperature, the more aromatic substances will be destroyed (due to overcooking). For this reason, and also to avoid the risk of an explosion, a still should always be run completely unpressurized. To avoid the effects of rectification, the column should be as short and wide as possible and the lyne arm should run downward at an incline.

Another vital part of a still is the steam thermometer. Without such a thermometer, the heads and tails can only be inaccurately separated by measuring the current alcohol content of the outflowing distillate, or, worse still, by the taste. We've suggested the steam-temperature method since the beginning of the 1990s; in 2011 Otfried Jung (Institut für Lebensmittelwissenschaft und Biotechnologie, FG Gärungstechnologie mit Lehrbrennerei,

Stuttgart-Hohenheim) confirmed this in an article published in a German professional journal. According to this article, the separation of the tails is much easier, and safer, if a steam thermometer is used. He also considered separating the tails by measuring the current alcohol content of the outflowing distillate an out-of-date method.

As we will describe in detail later in the chapter on distilling, the boiling point and the steam temperature are dependent on alcohol content. The lower the alcohol content in the kettle, the higher the boiling point will be. Knowing this, you can continuously measure the alcohol content of the distillate while distillation is ongoing, just like you would measure it after it's finished cooling with a hydrometer and other appropriate instruments, as is standard in industrial distilleries.

Unfortunately, the thermometer is often attached to the wrong part of the still. The alcohol content of the steam, and therefore the steam temperature as well, also depend on how high you perform your measurements: only the temperature at the highest level reached by the steam corresponds to the distillate's alcohol content. But where is this level? It is not the top side of the lyne arm, column, or still-head. The steam doesn't fill this part of the still up to the top because there is no over pressure. So, when it begins to boil in the kettle, the steam will rise, like water, until it finds an outlet. Then the steam will flow over the lowest edge of the outlet, never "touching" the top side. After flowing over this edge, the steam begins to cool down. Thus, in conventional stills, the correct measuring point is the "hot side" of the lower edge of the highest part of the lyne arm (see picture on page 47). This is the only location where the correct steam temperature can be measured. Measuring it at a lower point will give you a result that is far too high: the difference can be as much as twenty degrees Fahrenheit (ten degrees Celsius) or more.

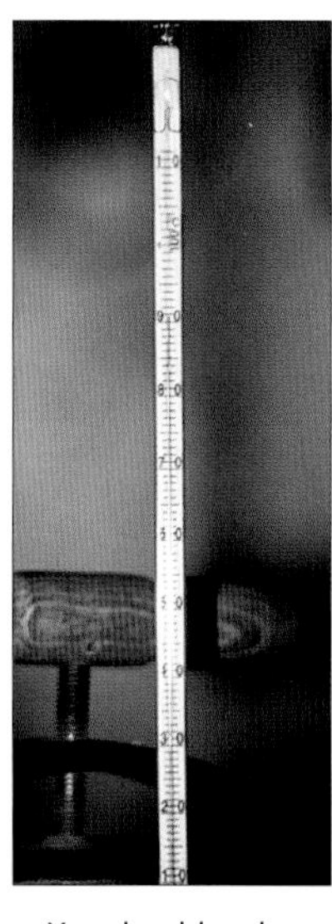

▲ You should make sure that you can read your thermometer at the full-degree scale Fahrenheit (half-degree scale Celsius). It's sufficient if it's effortlessly possible to estimate a half-degree Celsius between two tick marks.

The temperature of the steam at the point described is completely independent of the still's type or construction, so measurements taken here can be compared with measurements taken from other stills or described in books. The measurement sensor must be surrounded by steam and not be located in a dead spot. If the lyne arm comes directly out of the kettle, the thermometer can also be placed at the top of the kettle. The correct location for both variants is shown in the picture on page 49.

Due to its high accuracy, you should be sure to use a glass rod thermometer that you can read at the full-degree scale Fahrenheit (half-degree scale Celsius). It's sufficient if it's possible to effortlessly estimate a half-degree Celsius between two tick marks. The thermometer should go no higher than 230°F (110°C), as a higher range of measurement leads to less exact values. A standard rod thermometer with colored fluid works well since the color makes it easier to read. Round, bimetallic thermometers (furnace thermostats), used in many stills, have proven unsuitable because they are too inexact. Digital thermometers are only ever as exact as the temperature measurement elements attached to them; the many decimal places only feign a higher precision. The exact range of the measurement element should ideally be about 122°F–230°F (50°C–110°C). Digital thermometers require electricity and therefore batteries. The displayed temperature depends on the voltage, and this unfortunately depends on the battery level. Even half-full batteries display temperatures approximately six degrees Fahrenheit (approximately three degrees Celsius) higher than the true value, so we use the "old-fashioned" glass thermometers only.

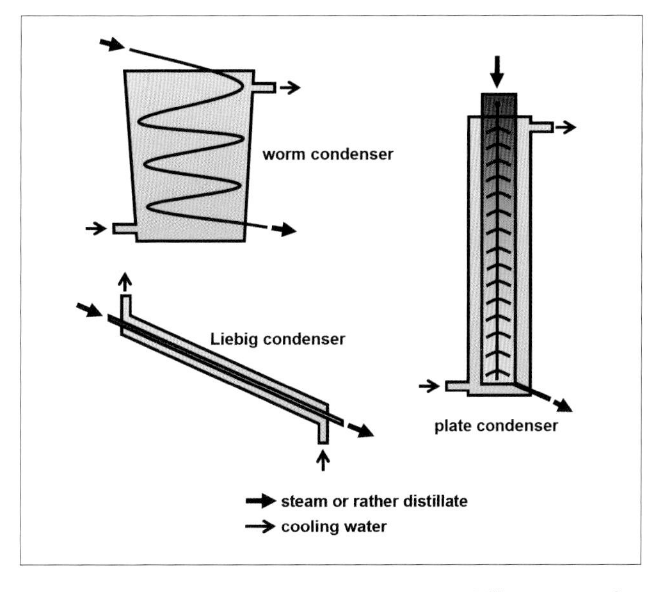

▲ Different types of condensers

Cooling

After the lyne arm, the steam condenses in the condenser. It is here that the distillate (i.e., the schnapps) is formed. So-called worm condensers are very popular due to their good cooling capability. This kind of condenser is made up of a container for cooling water with a spiral-shaped serpentine pipe (the worm) inside of it. The spiral is attached vertically such that the steam from the lyne arm enters the highest part of the worm, is cooled as it moves downward, condenses, and leaves the lowest point of the pipe as a liquid.

In most stills, the interior diameter of the worm is less than that of the lyne arm. Since the specific volume of the liquid is many times less than that of the steam, the liquid requires much less space. However, the worm must not be too narrow or else the condenser will be

Complete still with a condenser ▾

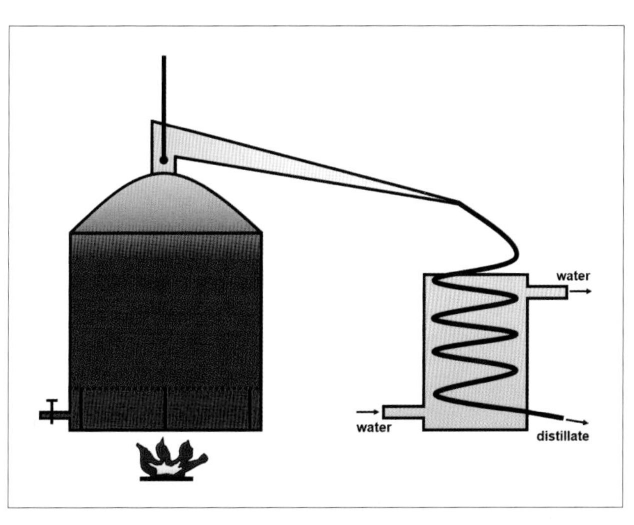

flooded during distillation and there will be a buildup of excess pressure in the still.

The length of the worm and the volume of the cooling container are the deciding factors for the condenser's cooling capability. Especially when working with smaller stills, the problem of the condenser being too small often comes up. Even if such a condenser is connected to a water line, it will not be able to fully condense the steam. The consequence: instead of schnapps dripping out of the condenser, only steam comes out. Even in small stills with a kettle volume of about half a gallon (2 liters), the worm should be at least 3 feet (a meter) long if the cooling container has a volume of about 1 quart. The spiral should continue uninterruptedly to the bottom, without any dents, or else the still will spit during distillation.

An alternative to the type of condenser described above for small stills is the Liebig condenser, named for its inventor, Justus Baron von Liebig. It makes use of a straight, encased pipe that leads downward at an angle. This type of condenser is relatively easy to clean, but it requires running water to operate.

Large stills are often equipped with a plate condenser. Like the worm condenser, this condenser also operates vertically. Its cooling power is not as great as the other two types of condensers, but it's easy to clean because the inner part, the frame with the plates, can be removed. Stills usually use a condenser made of copper or steel. However, you can also find glass condensers of various designs that can be connected to the distilling flask with gas-tight ground glass joints at laboratory suppliers.

Different types of stills

There are two different methods of distillation: simple distillation (pot still) and reflux distillation. In simple distillation, the material to be distilled is brought to a boil in a kettle and the resulting steam is immediately cooled back down (condensed). Distillation thus takes place in a single stage of heating and cooling. In a reflux still, a long, vertical pipe, the column, is located above the kettle. It condenses and revaporizes the steam several times consecutively, meaning it uses several distillation stages (see page 54).

These two methods produce two different types of results: in simple distillation, the alcohol and water are only partially separated, so, for example, a 12 percent ABV alcohol-water mixture (wine) could only be concentrated to an ABV of about 50 percent. This "imperfect" separation is an absolute necessity for producers of flavorful schnapps because a perfect separation leaves the flavor behind in the kettle. On the other hand, with reflux distillation, it's possible to produce more than 90 percent ABV alcohol from a 12 percent ABV wine in just one pass, albeit a more or less tasteless alcohol since the flavors will also be separated.

Schmickl's stills
0.5 up to 7.8 liters (pot still)

All of the following models are manufactured according to our engineering designs to meet the criteria listed under "Some key points." In order to meet point 6, all our stills are available with different sizes of the kettle. Thus, all parts of a still remain the same; only the kettle is different, and it can be changed easily. This system solves a problem distillers face regularly: having exactly the amount of mash required for the kettle size of his still. With the "classic" model—2 or 5 quarts (2 or 5 liters)—distilling with a spirit lamp takes about thirty minutes or an hour. Depending on the alcohol content of the material in the kettle, you can make 1.6–4 quarts (1.5–4 liters) of schnapps in one pass when using the 5-quart (5-liter) kettle. The "deluxe" model—2, 4, or 8.2 quarts (2, 3.8, or 7.8 liters)—uses a gas burner. Both the "vetro" and "piccolo" model—0.5 or 1 quart (0.5 or 1 liter)—also makes use of a spirit lamp.

Because of the steamer basket, it's not necessary to filter the mash to avoid burning it. This allows you to greatly cut down on flavor loss since the solid components (skins, fruit flesh) are generally very flavorful.

The steamer basket is also essential in producing spirits (see chapter 7). During distillation, the alcoholic steam removes the aromatic substances from the fruits or the herbs in the basket, giving you an excellent spirit.

The thermometer has a range of 32°F to 230°F (0°C to 110°C) and can be read to a precision of one degree Fahrenheit (half a degree Celsius). Measurements are taken precisely at the highest point that the not-yet-cooled steam reaches, giving you the proper steam temperature.

In the "classic," "deluxe," and "piccolo" models, cooling is carried out with a copper spiral and a copper container; in the "vetro" model, this part of the still is made of glass. Input and output nozzles allow you to attach a water line or a small circulating pump to any of the models. In the "classic," "deluxe," and "piccolo" models, the cooling container is open on the top, allowing you to fill it with ice water, for example, thus circulating cold water isn't absolutely necessary.

The stills can be used either for distilling mashes, beer, wine, and infused alcohol or for producing spirits and essential oils (see chapters 7 and 8). They can be obtained exclusively at our online shop, or directly at our house (see the end of the book for the address).

0.5 and 1 quart (liter) Schmickl still, "Vetro" model ▸

0.5 and 1 quart (liter) Schmickl still, "Piccolo" model ▸

◂ 2 and 5 quarts (liter) Schmickl still, "Classic" model

◂ 2, 4, and 8.2 quarts (2, 3.8, and 7.8 liter) Schmickl still, "Deluxe" model

Reflux stills

As we mentioned at the beginning, a reflux still is a device in which several distillations take place one after another. The difference to a pot-still is the column, connecting the kettle and condenser. The simplest kind of a column is a narrow, hollow, vertical pipe. The steam rises through the column, condenses along its sides after a couple of centimeters, then is evaporated again by the hot steam

Construction and installation of a steamer basket for distilling spirits ▸

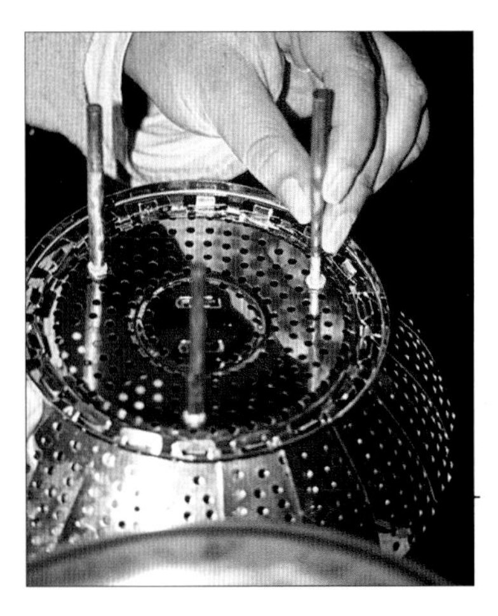

that follows it before condensing again a few centimeters higher, where it's evaporated again, then condensed again, etc. Each cycle of evaporation and condensation is in fact one instance of simple distillation. By the time the steam reaches the top, it will have gone through many of these cycles and thus been distilled that many times. Because evaporation requires energy and condensation provides it, a column works best if there is an energy equilibrium between both processes. Therefore, professional columns, as used in chemical laboratories, are vacuum isolated, like a thermos flask. In addition, they are operated with a valve at the top to adjust the reflux ratio (the ratio between the amount distillate led back into the column and the amount of distillate flowing to the outlet).

More distillations lead to a sharper and more exact separation of the alcohol and the water and therefore a more pure and concentrated distillate. If the column is high enough, tasteless 96 percent ABV alcohol will exit the still. As this is the azeotropic point (see page 60), the alcohol cannot be concentrated any further via distillation.

How reflux works ▾

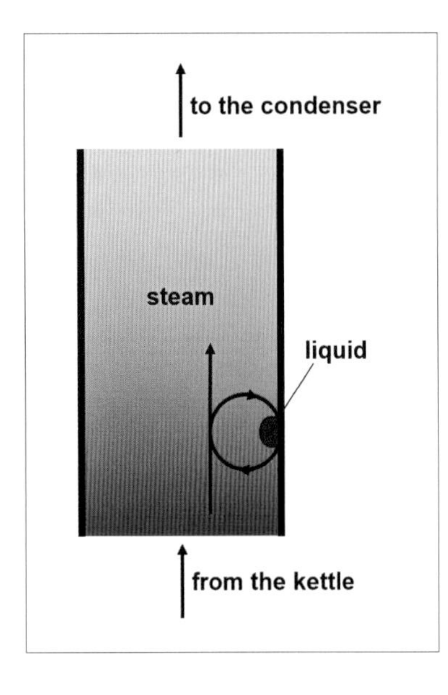

The cycle that takes place in a reflux still (evaporation, cooling, evaporation) is also called a "theoretical plate." The more theoretical plates in the still, the greater its separating ability. If the column contains filler materials, for example, Raschig rings (short, small tubes made of glass), Wilson-spirals (short, small spiral springs), nuts, steel or copper wool, packaging (e.g., thin, rigid metal grids stacked on top of each other), or other internal fittings, the separating ability per foot (meter) of the column will be increased. Some professional columns use what are called bubble cap trays, allowing large distilleries to distill their "double-distilled" products in just one pass.

In some areas, the general thinking hasn't yet changed to the idea that clear, high-quality fruit brandy produced with great effort shouldn't be used to get drunk as quickly as possible but rather enjoyed like wine. Therefore, stills that produce tasteless alcohol are common in some countries, and the alcohol produced from them is subsequently mixed with artificial flavorings.

However, we want to maintain as much (natural) flavor in the schnapps as possible, so these reflux stills are not suitable for our purposes.

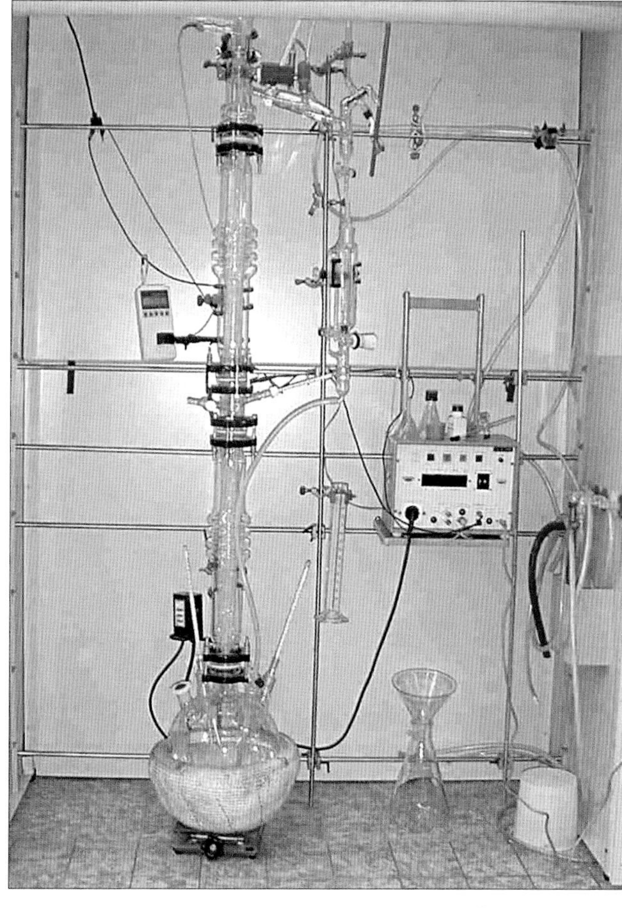

▲ Reflux still

Combination stills

Professional schnapps stills can vary from pot-type to reflux-type. According to one medal-winning distiller, the best brandies are made with as few theoretical plates as possible.

Large stills (pot still and reflux)

The Scottish whisky distilleries predominantly make use of simple distillation, as opposed to the large rum distilleries in the Caribbean, which use reflux stills.

Using the still

Filling

You should never fill the kettle all the way to the top. It would overflow and possibly also clog the column, the lyne arm, or the condenser. If this happens anyway, stop the distillation immediately and thoroughly clean the still. There is a risk of an explosion because even a small piece of fruit can be enough to clog the still.

If you're working with a steamer basket, put it into the kettle first and then pour in the (unfiltered) mash.

▲ Never fill the kettle all the way to the top with mash.

Combination still with agitator and jacket heating (Example from Arnold Holstein, Germany) ▸

Simple distillation in Scotland (bottom right); reflux distillation in the Caribbean (bottom left) ▾

☺ Tip: A steamer basket isn't always enough on its own if you're working with very pulpy mashes, such as apple or apricot. In these cases, place a paper towel over the basket before pouring in the mash.

It doesn't matter if there's still undissolved sugar left in the mash. It cannot evaporate, so it will remain in the kettle during distillation.

Receiver, container for the distillate

The receiver for the schnapps should always be set a little apart from the condenser's outlet so that the still will never be a completely closed system and a pressure equalization with the surrounding environment will be guaranteed. People often wrongly attach a hose to the condenser's outlet and stick it into a bottle so that "none of the distillate will evaporate." This is dangerous because it can lead to a buildup of excessive pressure in the apparatus, which, worst-case scenario, could cause an explosion. Additionally, the quality of the schnapps improves if it comes into contact with the air. None of the distillate will evaporate as long as it is cooled properly. The distillate must always be cold, not lukewarm. If it is lukewarm, you need to let more cold water circulate or the cooling will not take place to a great enough extent.

▲ Changing the receiver container

Emptying and cleaning

After each time that you use the still, the kettle should be thoroughly cleaned with water. Black deposits and stains often form in the kettle (we don't mean crusts or solid material), but their effect is only visual. If you have a copper still and wish to polish it, Brillo pads (small metal sponges covered with grease soap) work especially well. A descaler works very well for the worm. Citric acid dissolved in hot water also makes a great cleaning solution (about 3.5 ounces per quart or 100 grams per liter).

If you're using a new still, it's a good idea to thoroughly clean it before using it for the first time. First, clean it well with dish soap to get rid of leftover fat, oils, and metal shavings. Use a small brush with a flexible wire for hard-to-reach places. Or you can clean the still with acetone, since it mixes very well with water but also has strong fat-dissolving properties. Then wash out the acetone with dish soap and water. You can also clean the still with soda water or with a pulp made of cornmeal, salt, and vinegar.

In case of doubt, distill an old (bad) wine that you don't care about wasting as your first distillation. Some stills require ten or more test runs before they'll produce schnapps without a visible film of dissolved solder paste.

CHAPTER 4

Distilling

Introduction

In this chapter we will discuss how the distillate is produced.

Separating ethanol and water through distillation is only possible because of their different boiling points. Ethanol boils at 173°F (78.5°C) and water boils at 212°F (100°C). This means that at a pressure of one atmosphere (sea level), the temperature of a pot of water placed on a stove will continue to rise until it reaches 212°F (100°C). At that point the water will begin to boil and the temperature will not increase any further, regardless of how hot the stove is. Once all of the water has evaporated (the hotter the stove, the faster the water evaporates), the temperature of the pot, now empty of water, will continue to rise. The same is true of ethanol, but at a boiling point of 173°F (78.5°C).

If an ethanol-water mix is heated, it will not begin to boil at 173°F (78.5°C) (as pure ethanol would), but at a higher temperature. Just like a pure liquid, the mixture has its own particular boiling point, which will be somewhere between 173°F and 212°F (78.5°C and 100°C); the exact value depends on the relative amounts of water and alcohol. This means that the boiling point of the mixture depends on alcohol content. An ebullioscope makes use of this relationship. With this device, the alcohol content of the mixture can be determined from its boiling point. The advantage of this method is that dissolved sugar does not alter the measurement as much as it would with a hydrometer. Unfortunately, however, ebullioscopes are rather expensive.

Ebullioscope for determining alcohol content from the boiling point ▾

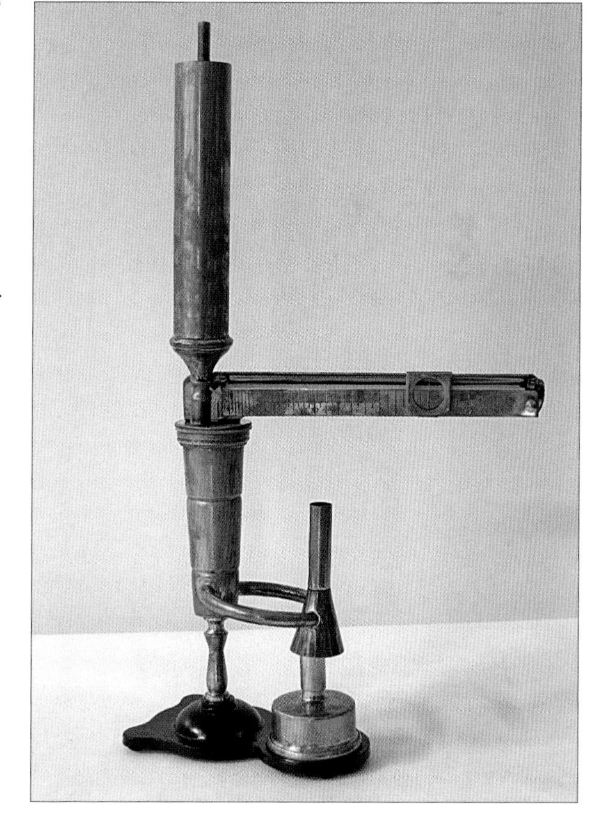

However, there is a difference between the behaviors of alcohol-water mixtures and pure liquids when boiling: as a rule, the alcohol content of the steam is not the

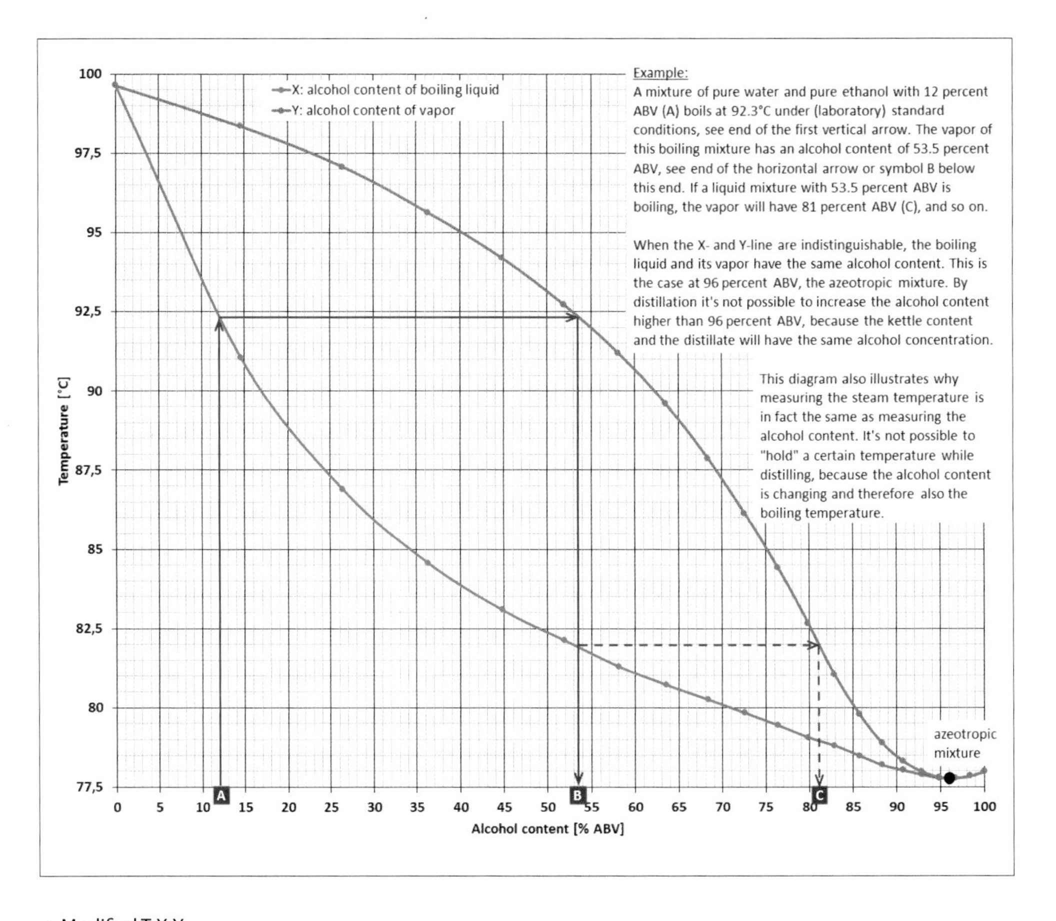

▲ Modified T-X-Y diagram: Common T-X-Y diagrams are plotted in mole fractions. We have converted those values into volume fractions for easier reading.

same as the alcohol content of the boiling mixture. Because the ethanol boils at a lower temperature, it accumulates more quickly in the steam, so while the steam is indeed also a mixture, it has a higher alcohol content. We make use of this effect in the distillation of schnapps: if the steam is cooled back down, the resulting condensate (the distillate or schnapps) has a higher alcohol content than the original mixture. This is possible up to an ABV of 96 percent, at which point the alcohol contents of the steam and the liquid are the same and no further separation through distillation is possible. Chemists call this the azeotropic point.

As a general principle, you should know that a bad mash can never result in good schnapps. Even the finest equipment in the world cannot save it. Only a good mash can give you good hearts!

The goal of distilling the mash is to separate the fruit flavors and the alcohol from everything else (decomposed and partially intact parts of the fruit, dead yeast, dyes, water, dissolved solids such as sugar, etc.), thereby concentrating the alcohol. This process produces three parts: the heads, hearts, and tails. The art is in effectively separating these from each other.

Heads

The head is the first thing that comes out of the still because it evaporates at the lowest temperature. It is produced through faulty fermentation. If you work under unclean conditions, heads will form from rotten fruit, stems, leaves, and other impurities. The cleaner your fruits and your methods, the fewer heads you'll get.

The cause of faulty fermentation and the resulting heads formation is unwanted microorganisms. There are many different kinds of bacteria and fungi, but only one of them produces alcohol. If the mash contains the wrong varieties due to rotten fruit or bacteria from the air, they will produce unwanted fermentation products such as acetaldehyde, ethyl acetate, butanol, acetone, acetic acid, butyric acid, hexanol, etc. Since these substances have a negative effect on taste and some of them are toxic, you should be sure to prevent them from forming. Aside from cleanliness, using a fermentation lock can significantly limit these microorganisms' reproduction. Adding cultured yeast also helps ensure that the "right" strain is the only one to multiply.

The word "methanol" inevitably comes up in the context of heads. As already described on pages 13 and 16, methanol is toxic and forms during fermentation due to ligneous impurities in the mash, among other things. It's impossible to keep a very small amount of methanol from forming in fruit mashes, but this is not a problem. Clean mashes contain a negligible amount of it. As we also described earlier, methanol and heads should not be confused. The heads do indeed include methanol, but if methanol forms in appreciable quantities in the mash, it will be distributed in all three parts, the heads, hearts, and tails. Schnapps with a high methanol content taste noticeably sharp, like pomace brandy, for example (see chapter 5).

Depending on the kind of distillate, the maximum allowed methanol content is defined between 35 and 53 ounces per 26 gallons (1000 and 1500 grams per 100 liters) of pure, 100 percent ethanol. Under normal conditions, if the fruit is fermented as already described in detail, you won't come close to reaching these limits. Sometimes you'll find news reports on people dying after drinking high-proof beverages. Almost without exception, all these cases are caused by stolen industrial-produced methanol, which is served pure or mixed with moonshine. The thieves usually don't know the difference between ethanol and methanol, and how dangerous the latter is. The antitoxin of methanol poisoning is interestingly ethanol, so such patients are given ethanol via drip-feeding. I know this sounds funny, but it isn't.

Methanol cannot be analyzed in the presence of ethanol directly, only via chemical reactions. One possibility for analysis is the Denigés' method: the oxidation of methanol with potassium permanganate to formaldehyde and its detection with Schiff reagent (fuchsin dye). Based on this method, a complete analysis set for home distillers and small commercial distilleries is available at Kuebler-Alfermi GmbH (online store Leo Kuebler GmbH, http://www.leo-kuebler.de/home_english.html, "Vinoquant 13 – Methanoltest," Art.Nr.: K6500).

Most of the heads is not made up of methanol, however, but rather of ethyl acetate, among other things. This substance is also what gives heads their unmistakable glue smell, because this solvent is also used in the paint and glue industries. Another significant component of heads is acetaldehyde. Because it's an intermediate product of alcoholic fermentation, it breaks down completely if the mash is stored long enough.

You can determine whether a distillate contains heads with a quick, easy-to-handle test sans laboratory equipment: the Pieper's first runnings separation test by Schliessmann Kellerei-Chemie GmbH (http://www.c-schliessmann.de/englisch/Startseite_englisch.htm). It will indicate how much acetaldehyde is present through a color change. It involves mixing 5 milliliters of the distillate with chemicals from three vials. The mixture will take on a color that you can match up with a color chart to identify any impurities and their concentrations (see page 67). You cannot use this test if you're distilling sulfurized wine because it causes the wrong colors to appear.

Conclusion: Working under clean conditions, cultured yeast, acid protection, a fermentation lock, and storing the mash can completely eliminate or minimize heads.

The alcohol content will be very high at the beginning of the distillation, usually over 80 percent ABV, which causes the first part of the distillate to smell pungent. Don't confuse this with the typical glue smell of heads! If you're unsure which it is, dilute a sample with water at a ratio somewhere close to 1:1, which should make it stop stinging your nose and cause the glue smell—if that is indeed what it is—to become a lot more noticeable.

You cannot use the heads themselves because they contain too many toxic substances. They are useful for cleaning windows, however.

Hearts

The next part, the hearts, contains the alcohol you're trying to produce, that is, ethanol plus whatever flavorful and aromatic substances were in the mash. The section after next will describe how to easily separate the hearts from the heads. If you used a clean, high-grade mash, the hearts will be the largest part by far and have an alcohol content of about 55–57 percent ABV.

Tails

The tails are not toxic. This part is made up of products that were broken down due to the lengthy boiling, which is why the tails can be made into a tasteless alcohol that is great for infusing.

This part also has a characteristic smell and taste: the distillate will stink like overcooked mulled wine and will taste feeble and bland compared to the fruity, aromatic hearts. The alcohol content will be very low by this point, about 20–30 percent ABV.

Distilling process

Because the boiling temperature rises continuously during alcoholic distillation, it's relatively easy to distinguish the heads, hearts, and tails from each other with the use of a steam thermometer. Do so as follows:

1. Turn the heat on high. Once the temperature begins to rise, reduce the heat by about half or three-quarters (i.e., between about 60° and 70°) shortly before (!) the first drop of distillate drips out from the condenser so that it doesn't start to boil too quickly. As we've already described on page 45, the mash must not boil too intensely if you want to create high-quality schnapps. After the first drop of the distillate appears, don't forget to adjust the heat to a level that causes the distillate to flow out at a moderate pace, (or depending on the size of your still) to quickly drip out of the condenser.
2. Although the first of the distillate is now dripping out from the condenser, the temperature is still increasing at a clearly perceptible pace. As long as this is the case, you'll get heads. Be sure to collect the heads in a separate container.
3. The temperature is remaining practically constant. If you observe the thermometer for a long time, the temperature will indeed continue to rise, but at an extremely slow pace relative to step 2. You are now producing hearts. Collect this part. The distillation should not be going too quickly. In a still with an approximately 2.5-gallon (10-liter) kettle, the distillate should drip out very quickly, but not in a continuous stream.
4. Once the temperature reaches 196°F (91°C), you must

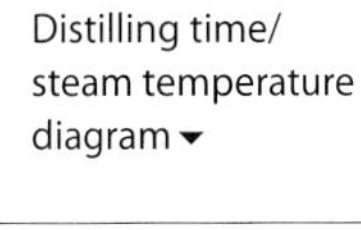

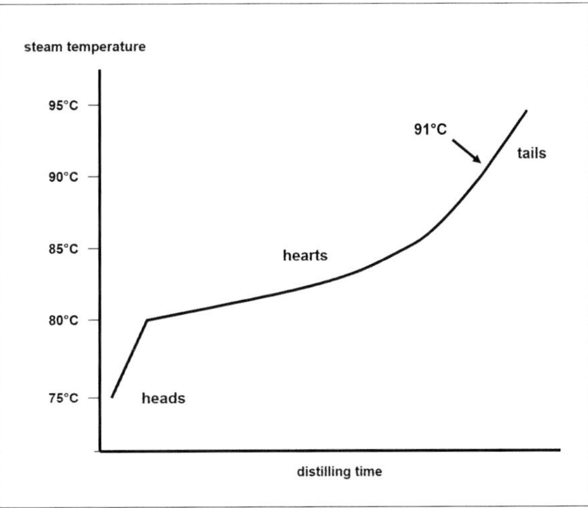

Distillation in pictures

▲ The steamer basket is placed in the kettle.

▲ Pouring in the mash. Never fill the kettle all the way to the top.

▲ Heating with a large flame.

▲ Reduce the heating power at about 149°F (65°C). Remove the heads until the temperature becomes practically constant.

▲ Change the receiver container and collect the hearts.

▲ The distillate should drip out quickly, but it shouldn't run.

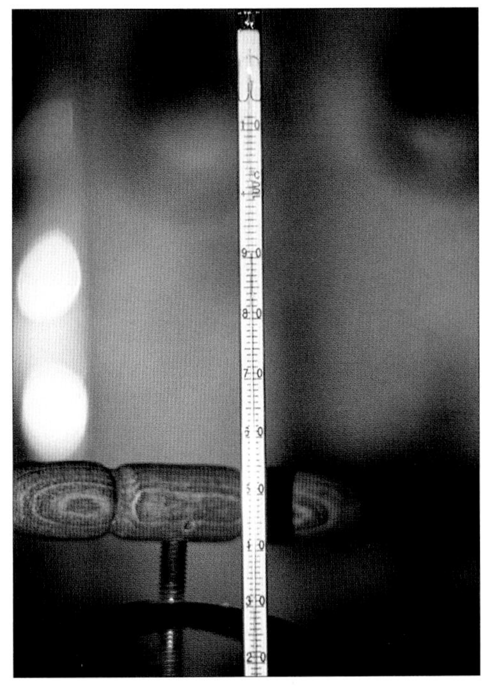

▲ Starting at 195.8°F (91°C), collect the tails in another container.

▲ The diluted distillate.

Note: If you're not sure whether you separated the heads, hearts, and tails properly, collect the distillate in many schnapps glasses and mark whether each one ought to be heads, hearts, or tails based on the steam temperature.
1. If the contents of the kettle have an ABV of 20 percent, the following is true. Up to about 176°F (80°C): heads; 176°F–195.8°F (80°C–91°C): hearts; more than 195.8°F (91°C): tails
2. Smell test: Can you perceive the smell of glue? Can you smell weak, boiled tails? Do the three stages correspond with the previous point?
3. Test the heads and the first part of the hearts with the heads separation test.

make sure to switch receiving containers. This is usually the point when the still starts producing tails. Collect the tails until you reach a temperature of about 199°F–201°F (93°C–94°C) because you can use it later.

The chart on page 63 shows how the steam temperature changes when you're using a high-grade mash. It increases very quickly until it reaches about 176°F (80°C) (step 2, heads), then there's a sharp change in the curve and the increase becomes very flat (step 3, hearts). Starting at about 190°F (88°C), the temperature increase accelerates somewhat again. From 196°F (91°C) onward, only tails will be produced.

Distilling store-bought white wine will not produce any heads because the way that it's made (the grapes are pressed and only the juice is fermented) keeps any from forming during fermentation. Tails will always appear and should always be separated. You will certainly notice one thing, however: the distillate will sting your nose almost like you're suffocating. This is due to the sulfurous acid that forms during wine production when sulfite is added. Almost every wine is "sulfurized" nowadays. A general rule of thumb is that the cheaper a wine is, the more it has been sulfurized. Fortunately, however, the sulfurous acid is very volatile and you can make the stinging smell disappear entirely vigorously mixing in air for approximately two minutes per quart (liter) with a hand blender (a hand mixer is not sufficient) or something similar when you dilute the distillate. A battery-driven cappuccino creamer or milk frother is perfect for quantities up to approximately 2 quarts (liters).

Relative proportions of the three parts

If you're using a 5-quart (5-liter) still and working under clean conditions, you should only get enough heads to cover the bottom schnapps glass at most (about fifty drops), certainly not more (for stills of other sizes, proportionately more or less). There's a widespread belief that you should separate out somewhat more to be safe, but this is a big mistake because the "choice cut" (i.e., the best of the hearts) appears immediately after the heads. If you do not include this part, the taste of your schnapps is sure to suffer.

If you want to use the tails, continue the distillation after you're through the hearts from 196°F (91°C) to about 199°F–201°F (93°C–94°C). This part will make up about a third of the total volume of the distillate. The following table gives you an overview of how much of the heads, hearts, and tails you can expect based on what you're distilling. These quantities only apply if you've worked under very clean and pure conditions.

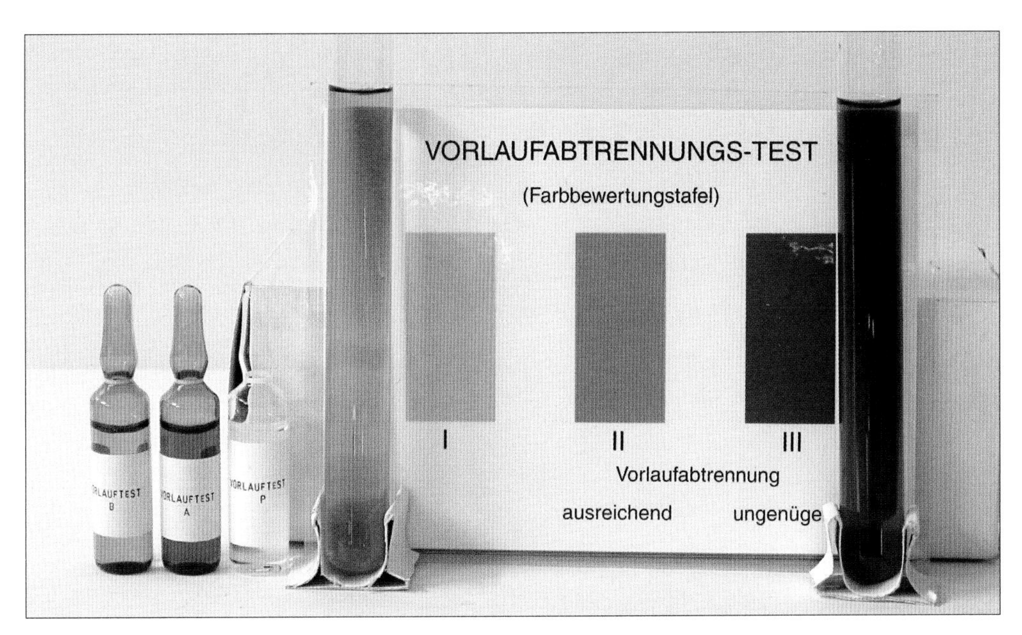

Pieper's first runnings separation test, from left to right:
Three vials with the testing chemicals
Test tube 1: test of a finished schnapps made from an eight-month-old, clean, high-grade mash. The mash was treated and the distillation was carried out according to the given instructions.
Color chart:
I, yellow: the sample is totally fine
II, light green, "adequate": the sample contains a limited amount of heads
III, dark green, "unacceptable": the sample contains a very large amount of heads
Test tube 2: test of a finished schnapps made through wild fermentation. The schnapps was double distilled with (insufficient) heads separation, but it still contained a large amount of heads. This color is unfortunately the most common in schnapps made this way.

Substance being distilled [unit of volume]†	Heads [unit of volume]	Hearts* [unit of volume]	Tails [unit of volume]
2000 of high-proof alcohol, 45 percent ABV (store bought)	0	about 2000	about 100
2000 of stored mash, 20 percent ABV	0.5	650–750	about 220
2000 of stored mash, 16 percent ABV	0.5	500–550	about 250
2000 of white wine, 12 percent ABV (store bought)	0	350–400	about 270
2000 of conventional mash, approximately 5 percent ABV	20-80	120–200	about 450

*Calculated to 43 percent ABV
†Any unit of volume, fluid ounces or milliliters

Relative proportions of heads, hearts, and tails

Substance being distilled	Start of hearts (°C)	Start of tails (°C)	Alcohol content of the hearts, undiluted (ABV)
High-proof alcohol, 45 percent ABV	79–79.5	91	70–73
Mash, 20 percent ABV	80–80.5	91	55–57
Mash, 16 percent ABV	80.5	91	51–52
Wine, 12 percent ABV	about 81	91	49–50
Mash, 5 percent ABV	>83	91	20–30*

*This is why conventional mashes have to be distilled twice.

Approximate values for the start of the hearts and its alcohol content after separation ▸

Antifoam

Some liquids, such as banana mashes or beer, tend to foam heavily when they're boiled. Even if you reduce the fill level in the kettle, it's hardly possible to prevent it from foaming over, which contaminates the lyne arm and the condenser. If the kettle foams over, you must end the distillation immediately and clean the still. For one thing, your distillate will be clouded, and for another, the situation is hazardous because the solid material can clog the still. It's even possible for the whole thing to explode. You should always end the distillation as soon as the distillate takes on the color of the mash.

Antifoam, an aqueous emulsion of polydimethylsiloxane (PDMS, also known as dimethicone, a certain kind of silicone oil, nontoxic and chemically inert), prevents foam from forming. As soon as you've poured the mash into the still, add about three to five drops of antifoam (for a 5-quart or 5-liter kettle) to keep it from foaming when it's boiled.

Note: Calibrating the temperature measurement: If distilling a substance of 12 percent ABV without any solids (e.g., wine), the entire collected distillate will have 51–52 percent ABV when tails begin. Even if your thermometer doesn't show 195.8°F (91°C) at this moment, the displayed value is the beginning of the tails for all distillations.

Single or double distillation?

Schnapps is often labeled as "double distilled." However, double distillation does not have anything to do with improving quality. On the contrary, each distillation causes some of the flavor to be lost because it makes the alcohol purer and more concentrated, and thus more tasteless. If you perform enough distillations in succession, you'll end up with a completely taste-neutral alcohol with an ABV of 96 percent—see the "reflux stills" section of chapter 3.

The only reason to do double distillation is an excessively low alcohol content in the distillate. If you're distilling something with an ABV of less than about 10 percent in a pot still, the distillate will have an ABV of less than 39 percent, meaning it will have to be distilled a second time. If what you're distilling has an alcohol content of greater than 10 percent, a single distillation is enough to reach schnapps strength (40–45 percent ABV) at least.

With conventional mashes—mashes without added sugar—the raw distillate (the result of the first distillation) will have an ABV of about 20–30 percent, depending on the fruit. A second run is necessary to exceed 45 percent ABV; the result is called fine distillate. To prevent losses in aroma and yield, you should produce as much raw distillate as is needed to fill the kettle up to at least three-quarters of its capacity. It's been proven that you lose less flavor if you only distill once and increase the alcohol content of what you're distilling to at least 10 percent ABV by adding high-proof alcohol in advance. For example mixing beer and vodka (the quality criterion of vodka is its purity, i.e., tastelessness) in a ratio of 78:22 to make a beer brandy or mixing the mash with raw distillate of the same fruit. With high-grade mashes, the first distillate should already have an ABV of about 55–57 percent, making a second distillation unnecessary. On the contrary, it will have to be diluted.

☺ Tip: If you're using a pot still . . .

. . . you have to perform two distillations if you're distilling something with an ABV under about 10 percent.

. . . you should only perform one distillation if you're distilling something with an ABV over about 10 percent.

If you do wish to perform a double distillation, do it as follows:

1. In the first distillate, separate the heads immediately and then distill up to about 199°F–201°F (93°C–94°C), that is, with tails.
2. The second distillate should therefore not contain any heads. Separate the tails as described at 196°F (91°C).

Calculating the amount when mixing a low-percentage alcohol with a high-percentage alcohol

As mentioned, in some cases the alcohol content has to be increased, for example when distilling beer with a pot still. Therefore a certain amount of alcohol with a low alcohol content has to be mixed with a certain amount of alcohol with a high alcohol content; thus the desired alcohol content must be between both of these contents.

AH = AL x (CL - CD) / (CD - CH)
AH: Amount of high-percentage alcohol [any unit of volume, e.g., fluid ounces or gallons (milliliters or liters)]
AL: Amount of low-percentage alcohol [same unit of volume as AH]
CH: Alcohol content of the high-percentage alcohol [percent ABV]
CL: Alcohol content of the low-percentage alcohol [percent ABV]
CD: Desired alcohol content of the mixture [percent ABV]

Example ▸ *We want to mix 5 gallons (liters) of beer, containing 4.5 percent ABV, with vodka, containing 39 percent ABV. The desired alcohol content of the mixture is 12 percent ABV. How much vodka do we need?*

Solution:
AL = 5 gallons (liters)
CH = 39 percent ABV
CL = 4.5 percent ABV
CD = 12 percent ABV
AH = 5 x (4.5 – 12) / (12 – 39) = 1.4 gallons (liters)
So a mixture of 5 gallons (liters) beer, containing 4.5 percent ABV, and 1.4 gallons (liters) vodka, containing 39 percent ABV, is in total 5 + 1.4 = 6.4 gallons (liters), containing 12 percent ABV.

Diluting to a drinkable strength

The hearts have a higher alcohol content than normal drinking strength so they have to be diluted. In general, this should be done immediately after distillation. The exception is if you're storing it in wooden barrels, because doing so causes some of the alcohol to evaporate, or if you're a commercial producer trying to keep storage costs as low as possible—the volume of undiluted alcohol is considerably smaller.

Determining the alcohol content

The alcohol content can be determined with a hydrometer and a transparent cylinder. A hydrometer is simply a floating object that allows you to measure the density of a liquid based on how deeply it is immersed in the liquid. Ethanol has a density of 0.8 kilograms per liter and water has a density of 1.0 kilogram per liter, so a mixture of the two will have a density somewhere in between those values. The higher the alcohol content of the mixture and, correspondingly, the lesser the density, the more deeply immersed the floating object will be.

You can buy specialized hydrometers called alcoholometers in specialty shops. They display the alcohol content directly in percent ABV. These devices are very precise—they're even used for calibrating—but the liquid must be free of dissolved substances, as depending on their concentration they can seriously alter the density. This means that dissolved sugar distorts the hydrometer measurements, and only pure distillates, which are sugar-free, can be measured accurately. This method will give you incorrect results with all mashes, wine, beer, liqueurs, infused alcohols, etc. Density also depends on temperature (the higher the temperature, the lower the density), so it's best to use an alcoholometer with integrated temperature correction, in other words an integrated thermometer. Most alcoholometers are calibrated to 68°F (20°C). If the temperature is higher, a correction value, given next to the temperature scale, must be subtracted from the measured value. If the temperature is less than 68°F (20°C), you'll have to add the correction value to the measured value.

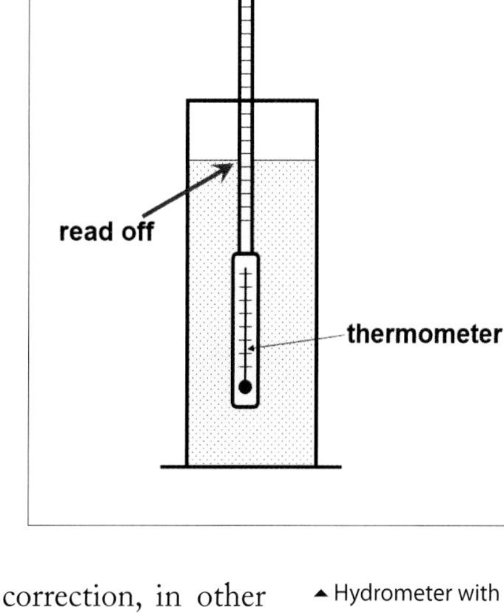

▲ Hydrometer with transparent cylinder

Fill the cylinder about three-quarters full of distillate, and then add the alcoholometer carefully. (Don't drop it; it's glass!) After thirty seconds to a minute, the temperature display and the immersion depth of the alcoholometer will have reached their proper values. For a correct measurement, make sure the hydrometer does not touch the cylinder wall or cylinder bottom. Usually, the percent ABV is read at an angle underneath the surface of the liquid (these are Tralles hydrometers), but check your alcoholometer's instructions for this information; others are read at an angle above.

Alcoholometers are available in different accuracies and ranges of measurement. Officials use huge devices with a measuring range between 35 and 45 percent ABV and an accuracy of +/- 0.1 percent or less. For home distiller purposes the much smaller ones, with a measuring range from 0 to 100 percent ABV and a tick mark of 1 percent ABV, are completely sufficient. You cannot taste the difference between, for example, 43.0 and 43.5 percent ABV. These alcoholometers have a total length of approximately 12 inches (30 centimeters). On the other hand, even shorter hydrometers, usually even without an integrated thermometer, are too inaccurate for our purposes.

Example 1 ▸ *The alcoholometer's display shows 65 percent ABV. The temperature is 68°F (20°C). What is the actual alcohol content?*
Solution:
The alcohol content is exactly 65 percent ABV because the alcoholometer is calibrated to 20°C.

Example 2 ▸ *The alcoholometer's display shows 65 percent ABV. The temperature is 77°F (25°C), and you see the number 1.9 next to this temperature. What is the actual alcohol content?*
Solution:
The alcohol content is 65 – 1.9 = 63.1 percent ABV.

Example 3 ▸ *The alcoholometer's display shows 65 percent ABV. The temperature is 50°F (10°C), and you see the number 3.5 next to this temperature. What is the actual alcohol content?*
Solution:
The alcohol content is 65 + 3.5 = 68.5 percent ABV.

Correction scale for alcoholometers ▸

Temperature of the liquid (°C)	Correction value for the measured value (% ABV)
5	+5.3
10	+3.5
15	+1.8
20	0.0
25	-1.9
30	-3.5

Calculating the amount of water needed for dilution

After this measurement, you now know the alcohol content and need to calculate how much water should be added to reach the final alcohol content you want. We always dilute to a drinking strength of 43 percent ABV, but if you prefer a different alcohol content, that's not a problem either. Spirits and brandies generally have alcohol contents between 40 and 45 percent ABV. At less than that you get a dull taste, and at more than that the tongue's taste sensors are numbed, making you unable to taste anything anymore. When you're deciding what alcohol content to choose, keep in mind that a low alcohol content can cause serious cloudiness.

You can calculate how much water you need to add as follows:

AW = AU x (CU-CD) / CD

AW	=	Amount of water to be added [any unit of volume]
AU	=	Amount of undiluted alcohol [same unit of volume as AW]
CU	=	Alcohol content of the undiluted alcohol [percent ABV]
CD	=	Desired alcohol content after dilution [percent ABV]

All amounts must be in the same unit, for example fluid ounces or gallons (milliliter or liter).

Contraction (meaning if you mix water and alcohol, the total volume is not the exact sum of their volumes, but somewhat lower) can be ignored for our purposes.

◂ Example 1

We have 500 fluid ounces (milliliters) of alcohol with an ABV of 63.3 percent. We want to dilute it to an ABV of 43 percent. How much water do we need?
Solution:
AU = 500 fluid ounces (milliliters)
CU = 63.3 percent ABV
CD = 43 percent ABV
AW = 500 x (63.3 – 43) / 43 = 236 fluid ounces (milliliters)
Thus, we must add 236 fluid ounces (milliliters) to reach an ABV of 43 percent.

Of course, after some brief algebra (that we'll spare you here), you can also use the dilution formula to calculate the amount of undiluted alcohol and the amount of water for the desired total amount of the mixture after dilution:

AW = AD x (1 – CD / CU)
AU = AD – AW
AD = Desired amount of diluted alcohol (all other variables have the same meaning as before)

◂ Example 2

We want to end up with 5 gallons (liters) of 43 percent ABV alcohol. We have 60 percent ABV alcohol available. How much 60 percent ABV alcohol and how much distilled water do we need?
Solution:
AD = 5 gallons (liters)
CD = 43 percent ABV
CU = 60 percent ABV

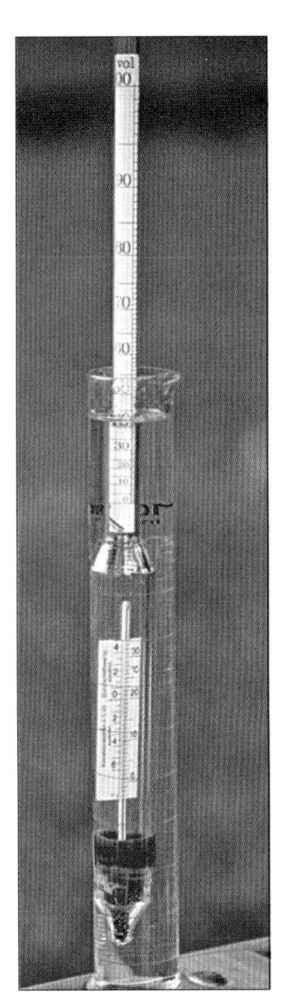

▲ Alcoholometer and graduated cylinder

AW = 5 x (1 – 43 / 60) = 1.42 gallons (liters) of water
AU = 5 – 1.42 = 3.58 gallons (liters) of 60 percent ABV alcohol
Mixed together, that gives us 5 gallons (liters) of 43 percent ABV alcohol.

Cloudiness

If you know how much water you need, carry out the dilution by vigorously stirring the alcohol while slowly pouring water into it. During this process, light to heavy milky cloudiness may appear due to the following causes:

- Mixing in tap water
- Different temperatures in the alcohol and water
- High concentrations of essential oils
- Excessive tails
- Distilling drupe mashes

Unless the water is very soft, you should not use tap water for dilution because the dissolved calcium salts are insoluble in alcohol and lead to snow-white flakes. It's better to use distilled water or deionized water, also known as demineralized water. Deionized water from a hardware store or supermarket, which can also be used for ironing or for car batteries, will do just fine. It's available in convenient gallon jugs, which usually cost less than five dollars.

☺ Tip: To check whether your groundwater or tap water is suitable for diluting, leave a diluted sample of your schnapps sitting for about two to three weeks at cellar temperature (i.e., cool but not cold). If no snow-white flakes appear after shaking the bottle at this time, your water can be used. Lots of high-quality fruit brandy producers advertise their products by stating that they only use their own spring water for dilution. The quality will doubtless be better than if using municipal water from plastic jugs, so you should make the test in any case.

Note: If you vigorously mix the brandy while you add the water so that it foams for at least three minutes per quart (liter), brandies from high-grade mashes will be ready to enjoy! See section on storage.

If the temperatures of the alcohol and the water are too different from each other, they won't mix well. In this case, leave them next to each other for a few hours and don't carry out the dilution until their temperatures have more or less equalized. This tip is often misunderstood. It is not necessary to wait until both liquids have exactly the same temperature. Only a large difference of more than approximately 40°F (20°C) can eventually cause a slight cloudiness.

When you're distilling something with a high essential oil content, such as skins of citrus fruits, pinecones, young shoots from coniferous trees, nuts, anise, or some other spices/herbs, the cloudiness is caused by the exact opposite of the issue with calciferous water. Essential oils are very soluble in alcohol, but not at all in water. This inevitably causes these distillates to take on a strong, milky cloudiness, similar to ouzo or Pernod, when water is added to them. If you increase their alcohol content again by adding undiluted alcohol, the cloudiness will disappear.

If the schnapps just has more of a gray "tint" than cloudiness, the cause is usually tails. The unpleasant aromatic substances typical of this stage of the distillation are not very soluble in water. Because of the low alcohol content of tails, even the undiluted distillate is often cloudy. If some of the tails are mixed with the hearts during distillation due to imprecise work, it will have to be distilled again or else it will be impossible to remove these aromas, which have a negative effect on quality.

Excellent drupe mashes can also cause a tint, but it is white, not gray—like three drops of milk in a glass of water. This kind of slight cloudiness appears while distilling the hearts, not just at the end when the alcohol content is comparatively low. In fact, its occurrence is a very positive sign, meaning everything, fermentation and distillation, was done perfectly and the product is rich in taste and smell. This effect doesn't appear if fermentation was less than optimal or if more aromatic substances than necessary were left behind during distillation.

☺ Tip: Be cautious if you see bluish or differently colored shiny cloudiness. This is caused by metals that were dissolved from the still, usually due to poorly made joints, green rust (verdigris), or low-quality materials. In this case, clean the still thoroughly before distilling the colored alcohol again.

As long as there is not an excessive amount of tails in the hearts, the cloudiness is just an (ugly) visual phenomenon that has no effect on quality or taste. The distillate should be clear, however. For that reason professionally produced, high-quality brandies are always filtered. There are special devices available to filter large amounts; for home distillers pre-folded paper filters (fluted filters) grade 1 are perfect. Stick one inside of another and filter the diluted (!) alcohol after a minimum of two weeks of storage, otherwise it will become cloudy again. Do not store the alcohol in the fridge or, even worse, the freezer before filtering. The lower the temperature, the more aromatic substances will clog and remain

▲ Pre-folded paper filters (fluted filters) grade 1

in the filter. If it's still cloudy after filtering, repeat the filtration process until the cloudiness has disappeared. You can dry the filters afterward and reuse them. For smaller quantities, classic coffee filter attachments make the best funnels. If the cloudiness is being caused by essential oils, you should note that each filtration takes away a small amount of essential oils and with it some flavor and aroma. In these cases, you have to weigh the taste against the visual presentation, although if the aroma is too intense or too pungent you may wish to diminish it somewhat.

☺ Tip: If you don't have a special filter, take two coffee filters and put absorbent cotton between them. In contrast to using grade-1 filters, leave the alcohol in the freezer overnight beforehand. This method sometimes works, but you'll need the special filter to deal with stubborn cases.

Treatment with activated carbon, taste-neutral alcohol

In the "tails" section, we described how the tails can be recycled. Collect the tails in a container and add about one heaping tablespoon of food-grade activated carbon for every 1.3 gallons (5 liters) of tails. Stir vigorously and let it infuse for at least forty-eight hours. The more time it has to work, the better the adsorption (i.e., the purification). You can also put the activated carbon in the container in advance and then collect the tails because the carbon works over a very long period (weeks or months). This results in the best purification.

In addition to the tails, you can also purify finished schnapps and wine (unwanted Christmas presents, often) in this manner as a way to clear out your cellar. This recycles the alcohol at least, allowing you to use it for future infusions. Don't use this method with viscous liqueurs: they will stick to the activated carbon, making the process useless.

After this process, put the alcohol with the carbon into the distilling kettle. You'll often read that you need to filter out the carbon first to keep the flavorful substances from dissolving again, but this is not true. The van der Waals forces that are responsible for the large molecules of flavorful substances binding to the carbon can easily survive temperatures of 212°F (100°C). You can of course filter it in advance if you wish, but, as described, it's not necessary.

During distillation, remove the heads first, if present (tails treated with activated carbon produce no heads), then collect the

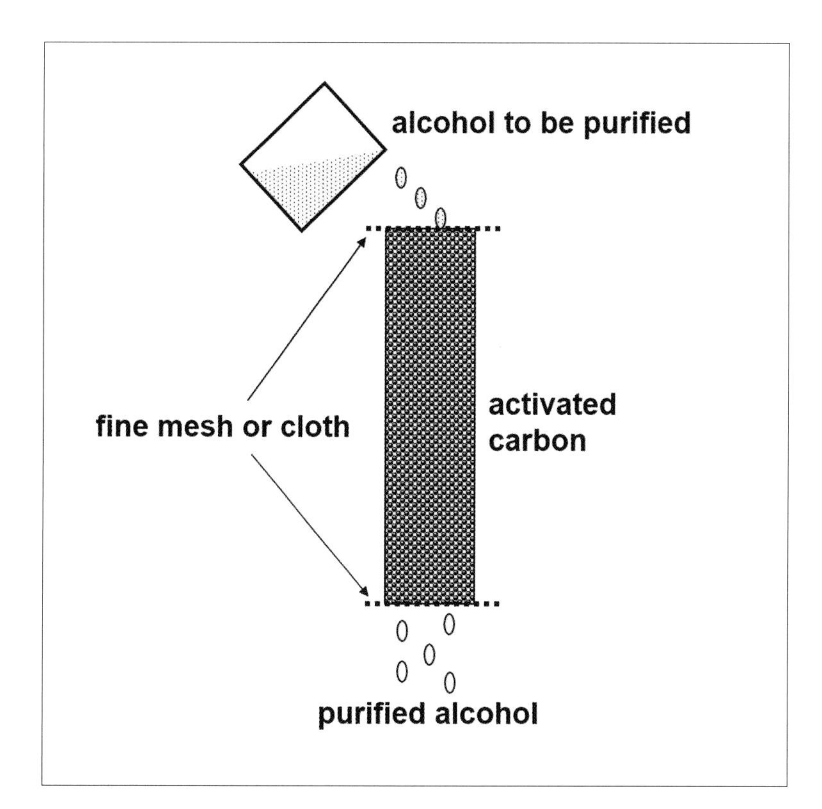

◂ Column filled with activated carbon

Note: You cannot make drinkable tasteless alcohol from denatured alcohol using activated carbon. This alcohol is intentionally poisoned with chemicals (denatured) that can only be removed through a complex chemical process.

hearts. You can discard the tails this time. The "hearts" will be a tasteless alcohol that is most suitable for infusing or spirits.

Another note regarding carbon: Only activated carbon (i.e., activated charcoal) is able to collect and adsorb flavors. Other types of coal cannot be used at all, not only because they have essentially zero adsorptive ability, but because they are sometimes highly toxic due to the sulfur and other impurities they can contain.

Another way of purifying alcohol using activated carbon involves using the so-called column method. In this case, you suspend activated carbon in alcohol and pour it into a pipe with an outlet valve. Make sure that no air bubbles are trapped inside. The column must never run while dry, so however much flows out at the bottom must be replaced with the same amount of alcohol that is to be purified at the top.

Storage

High-proof distillates should only be stored in glass or stainless steel containers. Pottery is permeable over decades, and wood will change the taste and color of the content (which can be desired), as described later. Be sure not to use plastic because the alcohol would gradually dissolve out chemicals that are hazardous to your

▲ Wooden barrels

health. You can store the distillate for a very long time without issues, if the container is tightly sealed and nothing can evaporate.

After a certain amount of time in contact with the air, the distillate becomes truly first class. To achieve this, seal the bottles with absorbent cotton and leave them standing in a cool environment for two to three weeks, at which point any pungent "sting" that may have been present will probably have evaporated and the taste will have become "rounder." For the same reason, professional fruit brandy producers store the distillates in airtight, sealed containers only filled to about two-thirds of their capacity for at least two but usually three to four years. This period is absolutely necessary because fresh schnapps made from conventional (not stored) mashes has an unenjoyable pungent taste and smell.

The following method has also proven effective: take a narrow silicone hose (not PVC!) and stick a piece of balsa wood into one of its ends. Weigh down the end with the wood using a metal nut and hang it into the bottle. With a small air pump (from an aquarium supplier), blow air into the free end of the hose and let it pass through the schnapps for about twenty-four hours at a cool ambient temperature.

You should always do the following procedure because it's easily done and in fact the same as stirring while you dilute your distillate: mix the fresh brandy for at least three minutes per quart (liter) with a mixer attachment for power drills, an electric hand blender (an electric mixer is not sufficient), or a milk frother vigorously enough that it foams (!) during the dilution. After this treatment, brandies from high-grade mashes are ready to enjoy!

Professional producers store the schnapps undiluted for economic reasons. We ourselves store all our distillates cool and diluted, hence readily available, in glass bottles up to a capacity of one gallon (five liters). When friends visit us, most of them want to try little samples of our treasures; this way it's not necessary to inconveniently dilute the distillates first.

We ourselves store all our distillates cool and diluted, hence readily available, in glass bottles up to a capacity of 1 gallon (5 liters). When friends visit us, most of them want to try little samples of our treasures; this way it's not necessary to inconveniently dilute the distillates first.

Storing the schnapps in bottles shows its quality. Only high-quality schnapps can be stored for a long period of time, whereas others would lose their flavor. We don't mean added flavor here, but rather just low-quality schnapps. This sort of schnapps will lose practically all of its flavor after being stored for a year, or serious taste defects will appear.

A tip for fans of whiskey, dark rum, grappa barrique, calvados, cognac, vieille prune, etc.: all these brandies are stored in wooden barrels for several years up to decades, mostly oak, often old, used (red) wine barrels. Other types of wood have become more and more popular because of their specific tastes. Due to the fact alcohol evaporates over time (the so-called angels' share), the distillate should not be diluted before storing it in a wooden barrel.

As a hobbyist, it's almost impossible to acquire this sort of barrel, but if the mountain won't come to Muhammad, Muhammad must go to the mountain. Just use, for example, oak chips or shavings, dosage approximately 0.2–2 ounces per 2.5 gallons (5–50 grams per 10 liters). Let the chips infuse in the schnapps for a few days to a few months, depending on the desired intensity, and it will take on a good oak-barrel flavor as well as a nice golden color. Of course, you can do this with any kind of distillate as well. Incidentally, this is exactly how cheaper whiskies and cognacs are made. Dark rum is stored in barrels that are toasted on the inside. This effect is also relatively easy to imitate. To do so, cook some wood shavings in a pan. As the shavings heat up, cover the pan with aluminum foil with holes poked in it. Oak chips can be ordered via the internet; several suppliers offer chips of different ligneous crops, untreated or in different degrees of roasting (e.g., toasted medium or dark).

It makes sense to infuse chips in red wine for a certain period of time before using them for distillates. In any case, fresh chips should be boiled for about ten minutes in water before using them for infusion.

Finally, you should consider the following: It isn't difficult to give a brandy a wooden taste, but it is an art to find the harmony between the fruit and the wood. The wood taste should enrich the original taste, not gloss over it.

CHAPTER 5

Mash Recipes

Excellent mash, excellent schnapps

Now that you have a solid base of knowledge on how to produce mashes, distilling, and the stills necessary for it, we are on the home stretch to enjoying schnapps. The following recipes will provide you with templates and suggestions.

Producing a mash is always done in a similar manner (as explained in chapter 2), so the core recipe that you should always stick to as a general rule is given again on the next page. Exceptions and idiosyncrasies are given in the descriptions for the individual fruits.

All of the ingredients and dosages in the recipes refer to a 26-gallon (100-liter) mash (of fruit pulp and water). If you're working with smaller or larger amounts, you can adjust the given quantities accordingly, but always use at least a heaping teaspoon of dry yeast mixture (e.g., turbo yeast).

As previously explained in chapter 2, you should never add more than one-third of the total amount in water. For a 26-gallon (100-liter) mash, that would be approximately 9 gallons (30 liters). The remaining 17 gallons (70 liters) should be fruit pulp. Chopping and mashing decreases the volume of the fruit, of course. Very soft, small fruits like plums, cherries, and grapes lose less volume than harder, larger types of fruit like quinces, apples, and pears. The values given in the table on page 83 are just general reference points, as they always depend on the ripeness, firmness, and, above all, size of the fruit. You can also use these values for fruits that aren't listed. The table shows how much fruit you need to get 17 gallons (70 liters) of fruit pulp.

◂ Grapes

Fruit pulp	Clean and wash the fruit, remove stems and rotten areas, mash them. With drupes, make sure that no pits are destroyed.
Water	As much water as is necessary to clean the equipment (at most a third of the fruit pulp). Exceptions are noted in the recipes.
Pectinase	Verflüssiger Spezial: 0.1–0.8* fluid ounces per 25 gallons (3–25* milliliters per 100 liters) of mash, if not already included in the yeast mixture
pH correction	Measure the pH value, correct it to 3.0–3.5 if necessary, e.g., with acid blend Biogen M: 10 fluid ounces per 25 gallons (300 milliliters per 100 liters) of mash
Yeast	High-grade mashes: Turbo yeast: 4 ounces (115 grams) per 21–26 gallons (80–100 liters) of mash, no additional nutrients needed. Sherry yeast + yeast nutrients: follow the manufacturer's given dosage. *or* Conventional mashes: All other types of yeast and nutrients or premade mixtures: check the instructions.
Added sugar (only for high-grade mashes)	Immediately: with turbo yeast: 29 pounds (13 kilograms), with sherry yeast: 23 pounds (10.5 kilograms) After about a week: turbo yeast: 29 pounds (13 kilograms), sherry yeast: 23 pounds (10.5 kilograms) After another week: turbo yeast: 29 pounds (13 kilograms), sherry yeast: 23 pounds (10.5 kilograms)
Fermentation	Let it ferment with a fermentation lock. Optimal temperature: 61°F–66°F (16°C–19°C). Stir once or twice a week. Stop stirring when the fruitcake has sunk; the fermentation is then finished (after about two to three months with high-grade mashes).
Storage	Conventional mashes: Store fully fermented mash at most one to two months (brandy at least two to three years); store high-grade mashes at least six months (no need to store brandy).
* The amount depends on the fruit. Soft fruit like drupes need less pectinase than hard fruit like quinces, rose hips, or rowanberries. Roots like Jerusalem artichokes need 0.8 fluid ounces per 25 gallons (25 milliliters per 100 liters).	

▲ Core recipe for a 26-gallon (100-liter) mash

Type	Weight		Volume of the whole fruits	
	in lbs.	in kg	in gallons	in liters
Apples	163	74	37	140
Apricots	157	71	30	112
Cherries	172	78	27	102
Prunes	172	78	28	105
Grapes	174	79	26	97
Peaches	165	75	32	123
Pears	170	77	31	117
Quinces	159	72	38	145

◂ Amounts of fruit for 18 gallons (70 liters) of fruit pulp

Agave (mescal, tequila)

Mexican agave forms the base for this distillate. Remove the large, succulent leaves and steam the remaining stem to produce a honey-like juice. Ferment this juice and add water at a 1:2 ratio. If you encounter problems during the dilution, proceed as with mead (see page 90). Organically produced agave syrup is also available in supermarkets, so you don't need to plant your own agave. Condenser is not necessary because there are no pieces of fruit in the juice.

Apples (Obstler, cider, calvados)

You don't need to remove the core, but you do need to remove the stems and any rotten areas. Add approximately 5 gallons (20 liters) of water per 21 gallons (80 liters) fruit pulp. Sour apple varieties don't have much flavor, so it's better to use full, aromatic types with yellow flesh.

Because of their subtly sweet taste, apples are very good for bulking up other types of fruit such as pears or plums. There's no

◂ Apples

fixed mixture ratio for the Austrian Obstler, usually a mixture of apples, pears, and prune plums. Distillers who make it tend to mix the fruit based on how much is available of each one, meaning that different Obstlers can have very different tastes.

If you distill the apple mash and add a small stick of cinnamon to the flavor basket, the taste of your apple brandy will have a touch of very nice, subtle, "Christmas punch" flavor.

The French calvados is in fact a cider brandy, stored in oak casks. For making cider, the apples are treated with a fruit press or juicer and the resulting juice is fermented without using the solid parts. Store the distilled cider for at least three to four years in oak barrels or use untreated oak chips as described at the end of chapter 4.

Apricots

Apricots produce a very fruity and flavorful schnapps, as long as the room temperature doesn't exceed 66°F (19°C) during fermentation. When using fermentation containers larger than approximately 40 gallons (150 liters), you will need a cooling device. Remove all the pits before making the mash to avoid an overly intense marzipan flavor. Due to apricots' high pH compared to other fruits with pits, it's important not to skip acidification.

Bananas

Bananas contain very little liquid of their own, so you should add water at a 1:1 ratio to keep the pulp from being too thick. Be sure to use overripe bananas; the peels should already be largely black. This makes the flavor especially intense. You must remove the peels before making the mash. Use antifoam during the distillation (see page 68) or else the banana mash will foam too much when it's boiled.

Beer brandy (bierbrand)

Beer brandy is a whiskey-like brandy produced by distilling beer. It's best to use beer from a supermarket. It's often possible to buy the dregs of barrels from breweries at a cheap price. Beer has to be double distilled for the distillate to reach a high enough alcohol content. Continue the first distillation until you reach about 201°F (94°C). Store-bought beer will not produce heads.

If, as already described, you add tasteless alcohol to the beer before distilling it so that the mixture has an ABV of about 11–13 percent, you'll only need to distill it once. Despite the addition of tasteless alcohol, the distillate will still have a stronger flavor than if you choose double distillation because the second stage takes away some of the flavor.

In contrast to every other type of distillate, the tails won't begin to appear at 196°F (91°C), but rather at about 197.5°F–198.5°F (92°C–92.5°C). You should collect the portion between 196°F–199°F (91°C–93°C) in small glasses to find the right temperature.

Note: If you distill beer, you should be sure to add a couple of drops of antifoam to the kettle or else it may boil over due to a large amount of foam forming.

The characteristic hoppy taste won't appear before 196°F (91°C) is reached.

Bock beer provides the highest possible alcohol and extract content, but everyone has his own preferences regarding taste, and there is a large variety to choose from among beers.

> ☺ Tip: You can put hops in the flavor basket to intensify the hoppy flavor of the bierbrand. For a 6-cup (1.5-liter) mash, add about 3.5 ounces (100 grams) of dried hops.

Blackberries

Blackberry wine is already a delicacy! However, it's hard to get the flavor into a distillate. For this reason, you should also place fresh blackberries in the flavor basket (see "Schmickl's small still" section).

Bullaces (Kriechenbrand)

The commonly wild grown black bullace is a variety of plum. The fruits are small, violet, round, very flavorful, and are particularly suitable for making high-quality, expensive drupe brandy with a flavor even more intense than with distilling a prune plum mash. Although all the following plum varieties are also clingstone cultivars (the flesh adheres to the stone) do not confuse the black bullaces with Mirabelle plums (round, yellow, smaller than Japanese plums), zibartes or cherry plums (round, yellow or red, same size as bullace), or damsons (ovoid, dark blue). Make the mash using also the undamaged pits. Add about 10 percent of the pits to the kettle during distillation.

Although often labeled as "damson brandy," the Austrian "Kriechenbrand" is in fact a bullace brandy. The English damsons have both a different typical flavor and shape compared to continental forms.

◂ Blackberries

Chequers

Chequers are picked when they're overripe, ideally after the first frost, but you should make sure—as with rowanberries—that the berries you pick haven't been gotten at by birds, who love to eat them. Picking them is a lot of work since they don't fall but must be picked by hand, and from trees that are usually 30 feet (10 meters) tall. Chequer brandy tastes similar to sloe brandy. Add about a third of the volume in water when making the mash to keep it from being too dry. Incidentally, the checker tree (*Sorbus torminalis* L.) was voted "Tree of the Year" for 2011 in Germany, and schnapps made from its berries is the most expensive kind of fruit brandy.

Cherries

Sour cherries have a particularly intense flavor, but normal cherries can make a good schnapps too. You should mash up the cherries well and leave the pits in; otherwise you'll lose the characteristic cherry brandy flavor. Try not to damage the pits, and remove any that do get broken. Add as little water as possible—this is especially important with cherries with regard to its effect on their flavor. During fermentation, the pits will stay in the fermentation container, and after it's finished they will collect at the bottom, completely separated from the fruit flesh. You can then easily fish them out with a sieve. Add about 10 percent of the pits to the kettle during distillation in order to maintain the characteristic cherry flavor. The fermentation temperature should be between 61°F (16°C) and a maximum of 66°F (19°C) or else the flavor will be "blown out."

Currants, red and black

Wash the berries with their stems and then remove the stems afterward; this is to avoid losing too much juice. Adding acid is not usually necessary. The fruit wine will already taste excellent. In the finished brandy, black currants will have a more powerful flavor than red ones, therefore mix both types if available.

Cherries ▸

Damsons

Damsons have been cultivated in England for centuries, so these plum varieties can be found in many English orchards and private gardens. They are also planted as hedging trees and windbreakers in certain parts of the country. A favorite of early English colonists in America, the tree has escaped from gardens and can be found growing wild in states such as Idaho.

Do not confuse it with the bullace or prune plum (see the corresponding sections). The damson is a clingstone cultivar (the flesh adheres to the stone) with yellow-green flesh and dark blue to indigo skin. Its shape is usually ovoid and slightly pointed at one end. The taste of the skin is very tart; the fruit is therefore not eaten fresh but used for cooking and making jam or other fruit preserves. It is perfectly suited for distilling; damson brandy has a considerably intense fruity drupe flavor. Make the mash using the undamaged pits as well as the fruit, and add about 10 percent of the pits to the kettle during distillation.

Elderberries

Water: Add about one-third of the volume of the berries (without their stems). You must remove the large, thick stems before making the mash. It's best to remove the stems from the berries entirely.

Elderflowers

Picking: Cut off the umbels and remove the thick stems. Do not wash or you'll lose the pollen! Because you can't wash them, it's best to pick the blossoms one to two days after it rains and far away from heavily trafficked roads.

You have to add a lot of water when using elderflowers because they don't contain any water of their own. Use one part water per one part softly squeezed elderflowers. Fermentation is only possible if you add sugar because the elderflowers don't contain any fermentable sugar. If you add orange and lemon slices directly to the mash, it will take on an especially refreshing flavor. For 26 gallons (100 liters) of mash, you should add about nine oranges and twelve lemons. Be certain to use untreated fruits with peels suitable for consumption, or else chemicals may enter the mash. If you decide not to add oranges and lemons, don't forget to keep an eye on the pH.

Gentian

Usually only the great yellow gentian is used in schnapps production, and even then only its root, which can be up to 3 feet (1 meter) long. But the red blooming gentian, the spotted gentian, and the brown gentian can also be used. A blue gentian is usually pictured on bottles of gentian brandy even though it's totally useless for schnapps production. Cut up the roots, which are harvested in the fall, and add water at a 1:1 ratio. Then continue according to the instructions. Use three to four times as much pectinase.

Most commercially available gentian brandy comes from apple mashes to which chopped gentian root was added at a 50:1 ratio in SI units (50 liters to 1 kilogram); this corresponds to 13 gallons to 2.2 pounds. Chopped, dried gentian root is available at distiller suppliers.

Grains (whiskey)

Grind the grain well and be sure to thoroughly remove any plant parts. This is a situation where you must follow the special instructions for starchy mashes (see chapter 2).

Only barley is used to make malt whiskey. After distillation, store the clear distillate in an oak barrel or in a carboy with oak chips (see pages 77–79 for a more detailed description).

Grapes (cognac, wine brandy)

Cognac is made using white grapes. Remove them from their stems and proceed according to the recipe. Store the clear, undiluted distillate in oak barrels if you have them available. If not, use carboys and add oak chips to the containers (see chapter 4, storage). After about four to five weeks of storage time, the cognac will have taken on its characteristic color. At this point it can be diluted to drinking strength (43 percent ABV). If you're storing it in oak barrels, you shouldn't dilute it until right before you bottle it because alcohol slowly evaporates in oak barrels. If it's already been diluted before being stored in the barrels, it's possible that your final product will have a lower alcohol content than cognac's typical 40 percent ABV due to this alcohol loss.

Grapes (grappa, pomace brandy)

Pomace is the name for the leftovers from the press in wine production—actually a waste product. If grapes are pressed and the juice is made into wine, the pomace can also be fermented and ultimately made into pomace brandy. Because of the high concentration of seeds and stems in pomace, pomace brandy contains a relatively high amount of methanol, causing it to taste unpleasantly sharp.

The following method produces an astonishingly flavorful and mild pomace brandy: Ferment Merlot pomace as a high-grade mash as described (use about twice as much pectinase as instructed), and after the second time you add sugar, but before the third, scoop off everything floating on the surface.

Grappa: First remove the grapes from their stems and then make them into a pulp. Ferment this pulp as usual. Winemakers pump out the grape juice that has accrued on the bottom and pour it back in at the top to ensure an even fermentation. However, we accomplish that in a more conventional way, by stirring. After the end of fermentation, winemakers pump out the fruit wine from the top and bottle it as wine. They lightly press the sediment underneath, making grappa. Pressing makes the fruitcake and the resulting grape juice bitter and gives the grappa a distinctive taste. Depending on preference, you can press it more or less strongly. The more strongly you press it, the bitterer the resulting product will be. There are many different production methods from this point:

- Use the unpressed fruitcake for distilling: Just press the fruitcake very lightly so that there's still enough liquid left in it.
- Press the fruitcake, dilute it with the pressed-out fruit juice, and distill: Depending on how hard you press, the wine will be more or less harsh, giving it the typical taste of grappa.
- Press the fruitcake, dilute it with the wine taken from the preceding wine fermentation process, and distill: Adding the wine from before instead of the pressed-out fruit juice makes the grappa milder.

The easiest method is to treat the grapes like a normal mash. If you want to intensify the bitter flavor of the grappa, press some of the grapes and then add them to the mash container. The unpressed mash (treat it like any other fruit) results in the highest possible quality and least sharpness in the distillate.

Jerusalem artichokes

Like sweet potatoes, this root contains both starch and fermentable sugar and can be fermented normally. Don't harvest Jerusalem artichokes until late spring of the following year; this will increase your yield and their flavor. You can add up to half the volume in water and a large amount of pectinase. If your pectinase doesn't give special instructions for Jerusalem artichokes, use three to four times the normal dosage. The taste is interestingly earthy.

Juniper berries (Borovička)

It's very rare to find pure juniper berry brandy from a mash nowadays. To make it, shake the berries from the bush in fall after the first frost and mash them. Add water at a 1:1 ratio because juniper berries are very low in water. Do not crush the berries, otherwise the distillate will get a strong resinous taste, like Swiss pine spirit.

◂ Jerusalem artichokes

The more common method is to make a mash out of apples and berries. Dried berries, available at spice traders, are a perfect alternative to the fresh ones. The uncrushed berries are added to the apple pulp before fermentation starts. Add about 2.2 pounds per 12 gallons (1 kilogram per 45 liters). This mixture tastes the same as the national brandy of Slovakia, the Borovička. Gin and Jenever are also based on juniper berries, but both are spirits, not brandies (see chapter 7).

Mangos

Skin the fruits and mash them up. You don't need to remove the fruit flesh from the pits because the yeast will do that during fermentation. After the pits have settled on the bottom of the fermentation container, separated from the fruit flesh, you should remove them and not use them in the distillation.

Maple syrup

The flavor of a maple brandy is very pleasant, although not as sweet as the syrup. You should taste the wine as well. Make the mash/wine like mead, see below. Organically produced original Canadian maple syrup is available in supermarkets.

Mead

You do not need to add sugar because honey is already about 75 percent sugar. Yeast cannot ferment something with such a high sugar concentration, so you have to dilute the honey 1:2 with water—2.2 pounds of honey and 2 quarts of water (1 kilogram of honey and 2 liters of water). This corresponds to about 0.9 pounds of sugar per quart (0.4 kilograms of sugar per liter), so if you're not using turbo yeast, you'll have to dilute it correspondingly more. Heat the honey and the water to 122°F (50°C) before you mix them. The temperature should definitely not be any higher than that or you'll destroy the honey's flavor. Stir it at 122°F (50°C) until all of the honey has dissolved in the water. Then cool the mixture back down to 75°F (24°C) and add all the ingredients as usual except for the sugar and the pectinase; instead of them, add 1 tablespoon (10 grams) of flour per quart (liter) of mash. The reason: every mash needs cloudy parts for the yeast fungus to adhere to. Since you have a clear liquid in this case, you need flour for this purpose. Mix the flour with cold water (so that it doesn't form lumps) and add this mixture to the mash. If you're using turbo yeast, you don't need to add any flour, like with a pure sugar-water solution.

Of course, the mead mash can be used not only for distilling, but also for making into mead.

Medlars

According to conventional wisdom, you should wait for the first frost before picking medlars so that they will be soft and easy to work with. But according to our experience, the inner part of the fruit is partly decomposed by then, causing a putrid smell in the distillate. So we harvest medlars earlier, at about the same time as apples, when the fruits begin to fall off the tree by themselves. Medlars are very solid, so you should add twice as much pectinase.

Nectarines

Remove the pits before you make the mash because it's very easy to damage nectarine pits, which can ruin the whole mash. Add only a little water in order to keep the flavor as intense as possible. The fermentation temperature should not be higher than 66°F (19°C) or you'll lose some of the flavor.

Nuts

Finely grind ripe walnuts and add about 1 quart (1 liter) of water per 2 pounds (1 kilogram) of ground nuts or else the mash will be too thick. The mash tends to foam during distillation, so you'll need to use a couple of drops of antifoam. Green nuts can be used as described in chapter 6.

Peaches

The same goes here as for nectarines: remove the pits (because peach pits also break open relatively easily), mash the fruit up with their skins, and use as little water as possible. This is also another case where the fermentation temperature should not exceed 66°F (19°C) under any circumstances. It is an art to produce a peach brandy with perceptible fruit taste and smell. We have noticed that it's crucial to use an appropriate variety of peach. The fruit should be extremely large, cultivated in a Mediterranean climate, and harvested as late as possible, when it is thoroughly very soft.

Pears

With pears, it's enough to remove the stems, leaves, and other impurities—do not remove the core or the seeds. If your pears are very hard, let them sit for a couple of days up to three weeks until they become soft before making them into a mash or it will result in schnapps with very little flavor.

◂ Pears

If you want to use different varieties of pears, such as perry (cider) pears and Williams' bon chrétien pears (known in North America as Bartlett pears), it's a good idea to mash them separately. Otherwise, you will not get the typical flavor of a specific variety.

Pineapples

You have to be careful if you choose to use this fruit. Imported pineapples are often barely ripe, making their flavor very sour and subdued. It doesn't even help to let them sit for a couple of days; they will decay from the inside out before the taste will improve. Instead, use canned fruit, which was harvested when it was really ripe. Sometimes there are also very small gourmet pineapples available, with a much better flavor. Of course, the best fruits are freshly picked locally. They result in an excellent pineapple mash.

Plums (vieille prune)

Make the mash using the undamaged pits; add about 10 percent of the pits to the kettle during distillation.

The "vieille prune," produced in Switzerland or France, is a plum brandy stored in oak barrels and coveted among connoisseurs. In Switzerland, the Swiss plum Löhrpflaume or Pflümli (round, yellow-red flesh, clingstone cultivar about the size of the Mirabelle plum, excellent eaten fresh) is used for the mash, and the distillate is stored in Swiss oak.

Potatoes (vodka)

You can put small amounts in a juicer and then mix the juice and solid parts together again. A chopper is ideal for larger amounts. Since potatoes contain starch instead of sugar, the mash will have to be made following the instructions for starchy products (see chapter 2). Make sure to use a type of potato with a high starch content, as this can vary between 9 and 30 percent.

Again, the quality criterion of vodka is its tastelessness; vodka was historically made of potato mash, which is triple distilled. Nowadays the mash is made of industrially produced starch. It doesn't make a difference anyway, when the distillate should be tasteless.

☺ Tip: If you steam the potatoes for about thirty-five minutes, you won't have to heat the mash water to 194°F (90°C). Add the amylase right after steaming them. Another advantage of steaming: the mash will not foam as much during distillation.

◂ The globular Japanese plums are the largest members of plums; prune plums (European plums) are smaller and oval, and the smallest ones, the globular black bullaces, give the most flavorful brandy.

Prune plum (slivovitz, Zwetschenbrand)

The prune plum is also known as the European plum or by its variety the Italian plum (ovoid, violet, freestone cultivar, waxy bloom, excellent eaten fresh). An absolute classic, and rightly so. Do not harvest the fruit before the skin is wizened. Mash the prunes with their pits, but make sure that they're not destroyed. Remove any broken pits. After the fermentation, the pits will have collected at the bottom of the fermentation container; you can fish them out with a sieve. It's a good idea to add about 10 percent of the (undamaged!) pits to the kettle during distillation—this is the only way to achieve the classic prune plum brandy flavor.

Although prune plum brandy is widespread among all central and eastern European countries, the national brandy of Serbia is the slivovitz variety, spelled šljivovica, which is often stored in oak barrels.

All German-speaking countries have uses the prune plum variety *Prunus domestica Hauszwetsche*—not to be confused with the Italian plum (*Prunus cocomilia*)—for centuries. Its origin is unknown. In comparison, the Italian variety does not taste as sweet, and the light-blue waxy bloom is a bit less intense than that of the Hauszwetsche. Therefore, a Zwetschenbrand (also spelled Zwetschgen- or Zwetschkenbrand) is not the same as slivovitz. High-quality Zwetschenbrand is an internationally renowned delicacy, with an extraordinarily smooth and fruity taste.

Quinces

Note: If the quinces are boiled for an hour or two until they're soft and brown before you make them into a mash, the flavor will be more intense.

Quinces are very hard and dry fruits. It's best to cut them up with a chopper or a juicer. Unfortunately, they're too hard to use rubber boots. The pulp will be very dry, so add about a third of its volume in water. Be certain to add pectinase or the skins and the fruit flesh won't disintegrate optimally. It's a good idea in this case to use twice as much pectinase as is indicated on the packaging. Interestingly, the small Japanese quinces are much more flavorful than the common large ones.

Quinces ▸

Raspberries As with strawberries, it's a lot of work to get enough raspberries to make a mash. But if you do manage to, you'll be able to make a very fruity mash with an intense flavor from them. Smaller amounts can be used very effectively to make infused liquor or spirits.

Rhubarb Cut the rhubarb into small pieces and make them into a pulp, leaving on the outer skin. It is very acidic and can be used effectively to acidify other mashes.

Rice (sake) There are two options for the fermentation here. The first is to proceed as usual for starchy products, working with raw, ground rice. The second option is to boil the rice, let it cool, and mix it with a small amount of water. Then add yeast, sugar, and a double dose of pectinase. This will also produce excellent rice brandy or sake.

Rose hips You don't need to remove the small pits for schnapps production, making it much less laborious than making jam from rose hips. You should cut off the black growths, however. It's best to pick the rose hips after the first frost. At this point they're fully ripe and very soft, making them perfect for mashing.

Rowanberries (Vogelbeerbrand) Rowanberry brandy or Vogelbeerbrand is sold for high prices in stores in Austria and northern Italy. If you pick the berries and then clean them thoroughly, you'll know why. It takes an enormous amount of work because the many small stems absolutely must be removed from the fruit. Some distillers recommend waiting for the first frost before picking them. We've been unable to follow this advice because birds, which are crazy for rowanberries, hardly leave any intact after the first frost. Because of this, we always pick rowanberries by the middle of October. The berries contain very little water, so you should add a third of the volume in water.

Sloes

Sloe brandy is treasured among connoisseurs. You should wait until the first frost to pick sloes. Mash them up with their pits, but make sure that they stay intact. After fermentation, collect the pits, which will have become separated from the fruit, from the bottom of the fermentation container. Use about 10 percent of the pits in the distillation.

You don't necessarily have to use fresh fruit for your mash. Dried sloes are available via the internet at distiller suppliers; 13 pounds (6 kilograms) of dried sloes per 12 gallons (45 liters) of apple mash is a perfect mixture. Add the dried sloes to the mash before fermentation starts.

Strawberries

Strawberries from a garden aren't very good for making into mashes because their flavor won't really come through. They're simply too watery. This is not the case with wild strawberries, which have an excellent aroma. But it's a huge amount of work to make a mash from wild strawberries because you need multiple quarts of them. It makes more sense to make infused liquor or spirits (see chapters 6 and 7). That allows you to achieve good results with just a small amount of strawberries.

Sugarcane (rum)

Press the sugarcane, ferment the juice, and then distill it. This is how you make white rum. If it's stored in charred oak barrels or treated with toasted oak chips (see chapter 4, storage), it will take on a mild, sweet flavor, at which point it's considered brown rum.

Sweet potatoes

This type of potato includes both sugar and starch. Because their sugar content is quite considerable, sweet potatoes can be fermented like normal fruit. Wash them, remove their skins, and mash them up well. Add up to half the total volume in water to keep the sweet potato mash from being too dry. Add about three to four times the dosage of pectinase.

Fruit wine

You don't always have to distill the mash—you can also use it to very easily make fruit wine, a truly organic wine free of chemicals and sulfurization. All you have to do is filter the mash and fine it.

Filtering the mash

Spread a metal or plastic mesh (e.g., a fly screen from a hardware store) over the opening of a large container, and make an indentation in it inside the container, forming a sort

Equipment for filtering the mash ▾

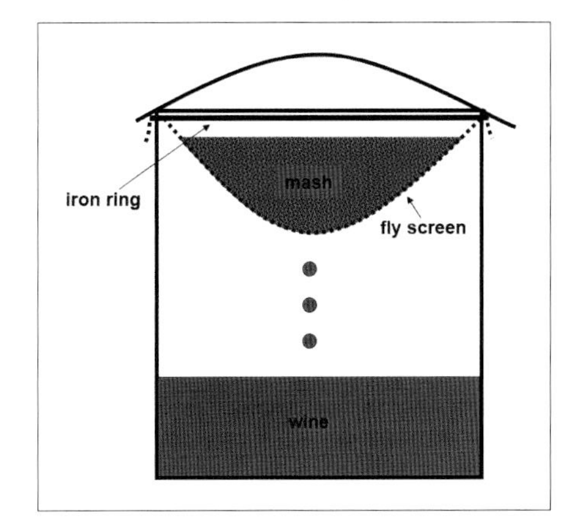

▲ Small fruit press

of sieve. Then secure the mesh with an iron ring, a clamp, or a wire around the edge of the container. Pour the mash onto the mesh, put a lid on top of it, and let it slowly drip through overnight.

Pressing the solid components

Collect the solid components caught by the filter and put them into a fruit press. If you don't have access to a fruit press, squeeze it tightly in a cloth. Do this by putting the solid components in the middle of the cloth, tying its four corners together, and turning it strongly around a stick.

▲ Squeezing out the fruitcake with a cloth

Extracting the fruit wine

Bottle the collected wine, seal the bottles, and let them sit for a few months. After this time has passed, the dead yeast and other cloudy particles will have settled at the bottom and there will be clear wine in the bottles. To extract the clear wine, stick a hose into a bottle, but not all the way to the bottom, or else you'll suck up the cloudy particles too. Place an empty bottle underneath and suck powerfully on the hose; the wine will then continue to flow on its own.

Storage

If you worked with a high-percentage mash, you can store the wine for a long time sans chemical additives without any problems because the mash is protected by its high alcohol content. However, you will notice when you drink it that your wine is stronger than many liqueurs.

In wines from conventional mashes, the alcohol content is too low to protect it from microorganisms, so it cannot be stored for a long period of time without adding preservatives (see chapter 2, storing the mash).

It's also true here that bottling the wine in glass bottles doesn't affect the wine, and storing it in wooden barrels affects it positively. You should avoid plastic containers unless they are made specifically for this purpose.

CHAPTER 6

Infusions

An infusion is the simplest method of schnapps production. It is accomplished by pickling fruits or herbs in tasteless alcohol. The alcohol extracts the aromatic and flavorful substances from whatever you're pickling, giving the infusion its taste and its color.

Materials

You'll need the following basic materials to make an infusion:

- Tasteless alcohol
- Distilled water
- Fruits or herbs
- A large pickle jar (1 gallon or 5 liters) with a lid

There are multiple ways to acquire the tasteless alcohol. One way is to buy liquor with very little flavor such as vodka, grain alcohol, or white rum. These spirits are made from potatoes, barley, or sugarcane and have alcohol contents of about 40 percent ABV.

A second option is neutral spirit. In some countries, pure, entirely taste-neutral alcohol with an ABV of 96 percent is available even in supermarkets. In others it is sold in pharmacies, sometimes regulated by the amount. An alternative is Everclear, a rectified spirit sold by the American spirits company Luxco, bottled at 75.5 or 95 percent ABV.

◂ Infusion

The third option is to make tasteless alcohol yourself using turbo yeast. Instructions for Prestige 8kg Turbo Alcohol Yeast:

Recipe for tasteless alcohol

> Pour 8 kilograms (18 pounds) sugar into 8 liters (2 UK gallons, 9 U.S. quarts) hot water, keep stirring until you are sure all the sugar has completely dissolved. Top up to 25 liters (5.5 UK gallons, 26.5 U.S. quarts) with cold water then stir for thirty seconds. Make sure the temperature is below 30°C (86°F) then add the sachet of yeast. Stir for one minute. For maximum alcohol and best quality, ferment for ten to fourteen days at between 19°C–23°C (66°F–73°F) air temperature. *Never* make more than 25 liters in one batch.

To avoid possible taste defects, such as the taste of yeast, it's a good idea to let the fully fermented sugar-water mash sit for about six months before distilling it, like a high-grade fruit mash. After distillation with a pot still, the alcohol content will be about 55–60 percent ABV.

A fourth option is to use tails that have been treated with activated carbon and distilled a second time. The alcohol content will depend on your products in this case, so you'll have to measure it yourself.

Depending on the strength of the tasteless alcohol you use, it may have to be diluted with distilled water (see page 73). If you're going to distill the infusion later, you can use tap water instead. There are two different alcohol contents used for infusions: 43 percent ABV for fruits that don't contain water and for herbs, and 53 percent ABV for watery fruits.

Infused alcohol	Infusion substance
43 percent ABV	• All types of herbs • Orange and lemon peels • Spruce shoots • Elderflowers • Coffee beans • Etc.
53 percent ABV	• Raspberries • Currants (red or black) • Sloes • Pinecones • Green, unripe nuts • Cherries • Wild strawberries • Etc.

As a general rule, the higher the alcohol content, the greater the extractive ability. Herbs that are infused into 96 percent ABV alcohol give the liquid a deep green tincture and become snow white themselves because the alcohol extracts all of their color. They also crumble like thin glass between your fingers. However, at 96 percent ABV, many substances are also dissolved out that don't necessarily contribute to a good flavor.

With fruits, it's best to use fresh ones as these contain the most flavor. But frozen fruit such as raspberries also work well for infusing. You should pay attention to one thing when choosing your fruit: that it is flavorful and not too watery. Cherries, wild strawberries, red or black currants, or sloes are thus perfect. Melons, home-grown strawberries, or kiwis contribute more of a "nothing" flavor when used in infusing. You should only use dried fruits if it's really the taste of the dried version of the fruit that you want, as this is very different from the fruit's original flavor. Before you place the fruits in the infusion container, you should thoroughly wash and clean them. Dirtiness or rotten fruit will irredeemably ruin the whole product.

☺ Tip: If you're working with store-bought spirits with an ABV around 40 percent, you should only infuse with products from the "43 percent ABV" list. Otherwise your infusion products will be too watery and will dilute the alcohol.

Store-bought neutral spirit (96 percent ABV) can be easily diluted to 48 percent ABV by mixing it with water at a 1:1 ratio. You can then choose from any of the infusion products.

You can grow herbs yourself or you can buy fresh ones from the supermarket. Pre-made herb mixtures are also available at distiller suppliers or pharmacies/drug stores. Again, the herbs must be cleaned well before you add them to the infusion container. You can also add the herbs to the infusion container gradually. If you filter out the herbs after using them in an infusion, you often can use them in another one later.

You can create a wide variety of enjoyable mixtures using herbs. Mixing fruits, on the other hand, results in an indefinite fruity taste, although it is excellent for rum pot, a traditional German dessert.

In principle, you can infuse with anything as long as it isn't toxic, including, for example, coconuts, coffee beans, garlic, morels, and many others. Be sure not to add any sugar or cream to the infusion. If you do, it will be considered a liqueur.

Note: Your container doesn't necessarily have to be a pickle jar. All that matters is that it has a large opening. It must not be made of plastic because high-percentage alcohol can dissolve some substances from plastic that not only hurt the flavor but are also toxic.

Fill the 1-gallon (5-liter) pickle jar about a third of the way full with fruits or herbs. Be careful if you're using resinous or bitter substances; adding too much can destroy the flavor. However, it often helps in this case to distill again. There are also some herbs and spices that you should be careful not to add too much of, including lovage, lemon peels, and juniper. If you add more than a third of the total volume in watery fruits, the alcohol that you're infusing will need to have an ABV higher than 53 percent because the fruits will add water to the container, diluting the alcohol.

After you have added your infusion substances, fill the container the rest of the way with alcohol and close it up airtight. You should leave the fruits/herbs inside for at least six to eight weeks to get a good flavor. It's best to store the infusion at room temperature and to put it in the sun every once in a while, which is especially beneficial. There's nothing wrong with leaving everything in the alcohol for longer, unless you're using berries. They should be removed after about six to seven weeks at the latest or else the bitterness of the many small seeds will slowly start to dissolve into the alcohol, making it bitter, too.

Infusions – preparation

Tasteless alcohol	Alcohol content: 43 percent ABV, 53 percent ABV if using watery fruits
Quantity of fruits/herbs	About a third of the container's volume
Infusion time	At least six to eight weeks (Six to seven weeks at most with berries)

Once the infusion is finished, you can drink it directly. The leftover fruits also make great desserts. But be careful, the high alcohol content will be present in the fruits, too. Infusions with resinous or bitter ingredients such as pine, nuts, citrus peels, etc., should be distilled; otherwise, they'll really only be palatable to "specialists."

If you prefer clear, colorless—but not tasteless—alcohol, you can distill the infusion without diluting it first. If your still includes a steamer basket, fill it with fruits/herbs from the infusion because they contain very large amounts of flavor and alcohol. If your still does not have a steamer basket, add the fruits directly to the liquid.

Making an infusion

◂ Chop the fruit into small pieces (green nuts in the picture).

▴ Fill the container a third of the way full with them (you can also use frozen items, like in the picture here).

▴ Dilute to 50 percent ABV with alcohol that doesn't have much flavor.

▴ Fill the infusion container with 5 quarts (5 liters) of alcohol.

▴ Let stand while sealed for at least six to eight weeks (left: fresh, right: after eight weeks).

Recipes

When deciding which ingredients to infuse with, you can let your imagination run wild—it's almost impossible to do something wrong. You can find the necessary alcohol contents for the different fruits on page 98. All quantities assume a 1-gallon (5-liter) container. Usually, the container should be filled a third of the way full with infusion ingredients and then filled the rest of the way with alcohol. Exceptions to this rule are noted in the recipes.

Absinthe

See wormwood.

Blackberries

Wild blackberries or blackberries from a garden are best to use. The blackberries available in stores don't usually contain enough flavor. Let the infusion sit for four to five weeks. Do not perform a subsequent distillation or none of the flavor will be left in the distillate. But if you do distill it with fresh berries in the steamer basket, the clear blackberry schnapps will have a good flavor.

Blueberries

It's best to use wild berries that you pick yourself because they have a much more intense flavor than the often enormous ones available in supermarkets. Let the infusion sit for at least five to six weeks. You should not distill it afterward or you'll lose a large amount of the flavor. Distilling is only an option if you place additional fresh berries in the steamer basket; this is the only way to give the distillate enough flavor.

Cherries

You should use very ripe cherries and make sure that they don't contain any maggots. Sour cherries work especially well because they usually don't contain any maggots and are especially flavorful. Don't fail to enjoy the fruits in a dessert! If you distill this infusion, it will lose too much of its flavor, unless you add fresh cherries to the steamer basket.

Chili peppers

Adding three to four red chili peppers to 1 gallon (5 liters) of tasteless alcohol will give you a truly spicy infusion. You can also use spicy jalapeños instead. You should definitely not let the infusion sit for more than two to three weeks or the schnapps will be too spicy to drink.

◂ Elderflowers

Coffee

Add 27 fluid ounces (800 milliliters) of coffee beans or 13.5 fluid ounces (400 milliliters) of ground coffee to the container. You can let it infuse for as long as you want. If you don't want to distill the infusion afterward, filter the ground coffee with a coffee filter.

Currants, red and black

Fill the container a third of the way full with the fruits. Remove the stems completely. Black currants provide a more flavorful infusion than red ones. Let the infusion sit for a maximum of one and half months—any longer and it will take on an unpleasant off taste.

Elderflowers

You can add elderflowers to the infusion container along with their thin stems. Fill the container about half full with them so that the taste comes through strongly. Adding citrus fruits (two oranges and one lemon per gallon or 5 liters) to the infusion gives it a particularly refreshing taste.

Gentian

Infuse with washed and cut-up roots of the yellow, red blooming, spotted, or brown gentian. About 50 grams of dried gentian per gallon (5 liters) is enough. Blue gentian is not suitable for schnapps production.

The root has a very intense flavor, enough so that you can use it for a second infusion afterward. It's a good idea to distill the infusion once.

Hazelnuts

Lightly roast the nuts before the infusion. This makes their flavor much more intense. Then cut the nuts in half and add them to the infusion.

Herbs

You can make herbal schnapps out of a countless variety of mixtures and variations. Our Ulrichsbergbitter (St. Ulrich mountain bitter) is a deep-brown herbal schnapps made from 1.9 ounces (54 grams) of wormwood, 0.7 ounces (20 grams) of lemon balm, 0.7 ounces (10 grams) of sage, 0.2 ounces (5 grams) of oswego tea, 0.2 ounces (5 grams) of althaea (marshmallow) leaves, 0.2 ounces (5 grams) of rue, 0.04 ounces (1 gram) of lovage, 0.2 ounces (5 grams) of mint, 0.2 ounces (5 grams) of southernwood, 10 petals of wolf's bane, and 0.2 ounces (5 grams) of clary per gallon (5-liter) batch. It becomes very dark after about one to two months due to the lemon balm. The distilled version is also excellent.

If you want to infuse with a single type of herb, we recommend the following: lemon balm, sage, rue (*Ruta graveolens*), or wormwood.

Hops

You can use wild hops, which can be found almost anywhere, for the infusion. You can leave the infusion sitting for as long as you want. After a short time, it will take on a red-brown color. If the taste is too spicy or bitter for you, just distill it once.

Lemons

Be sure to only use fruits with untreated (i.e., unwaxed) peels. Zest the fruit with a potato peeler—without the white part—and fill the container a third of the way full with loosely packed peels. Let it infuse for at least three to four weeks. It tastes great infused or distilled. This is another case where you are likely to run into cloudiness during dilution due to lemons' high essential oil content.

Limes

Using a potato peeler, thinly zest the limes (with untreated peels). As little of the white part of the peel should be used as possible to keep the infusion from becoming bitter. Very loosely fill the container a quarter to a third of the way full with the zest. You can let the infusion sit for as long as you want. This infusion is very wholesome, and the distillate is excellent, too. However, due to the essential oils, you should expect milky cloudiness when you dilute it.

Morels

Something for experts: pickling dried morels. Fill the container at most a third of the way full; that's all you need. You can let this mixture infuse for as long as you want. Incidentally, pickled morels look very interesting!

◂ Lemons, oranges, and limes

Mushrooms

It's hard to believe, but even (wild) mushrooms can be used in infusions. Fill the container a third of the way full with whole mushrooms. Mushrooms that produce especially interesting tastes include chanterelles, morels, parasol mushrooms, and penny buns. Warning: Ink caps are poisonous when combined with alcohol.

Nuts

Infused nut schnapps is a particular delicacy. The nuts should be harvested in early to mid-June, when they're still soft inside and haven't yet formed a hard shell. Nuts that are still green can be infused with the shell. Large nuts should be cut into quarters, and small nuts should be cut in half. For Johannistrunk, a traditional liquor in the alpine region, the nuts must be harvested on the feast day of St. John the Baptist (June 24), Honey, cinnamon, and cloves are also added.

Fill the infusion container a third of the way full. You can let it infuse for as long as you want, but it will already be completely black or a deep dark green after three to four weeks. Nut infusions are very bitter, almost unpalatable. They don't attain their good, drinkable flavor until after being distilled, because the bitter particles are left behind in the kettle.

☺ Tip: If you're working with green nuts, be sure to wear gloves! After we had finely cut up the nuts, our hands had become brown-black from their juice. Washing, rubbing, and cleaning didn't help. We had to wait until the color disappeared on its own (three to four weeks)!

Oranges

Winter is the best time for orange infusions because they are cheap and available everywhere. Be sure to use untreated oranges with unwaxed peels. Use a potato peeler to remove the outermost peel. Remove as little as possible of the white part of the peel because it contains the orange's bitterness. Fill the infusion container a quarter or at most a third of the way full with very loosely packed peels. After two to three weeks, an excellent orange schnapps will have formed. It tastes good undistilled or as a clear schnapps. If you choose to distill the infusion, expect cloudiness during the dilution because the peels contain a very large quantity of essential oils.

Raspberries

Wild raspberries are the best possible option here too. Raspberries are a very flavorful fruit, so you can also use frozen berries from the supermarket without having to do anything further to them. Again, keep an eye on the infusion time here: six to seven weeks at most. The "drunken" raspberries taste great with whipped cream!

> ☺ Tip: We bottled a clear raspberry schnapps in a nice bottle and added a couple of raspberries for decoration. It was going to be a gift. When we saw the bottle a few weeks later, we were horrified. The raspberries in the bottle were almost colorless and completely spongy. We therefore don't recommend adding decorative raspberries to your clear schnapps, as the fruits lose their color and appearance quickly.

Rose hips

Pick the rose hips after the first frost; this is when they have the best flavor and are already soft. Cleanly remove all leaves and stems and let the infusion sit for at least four to five weeks.

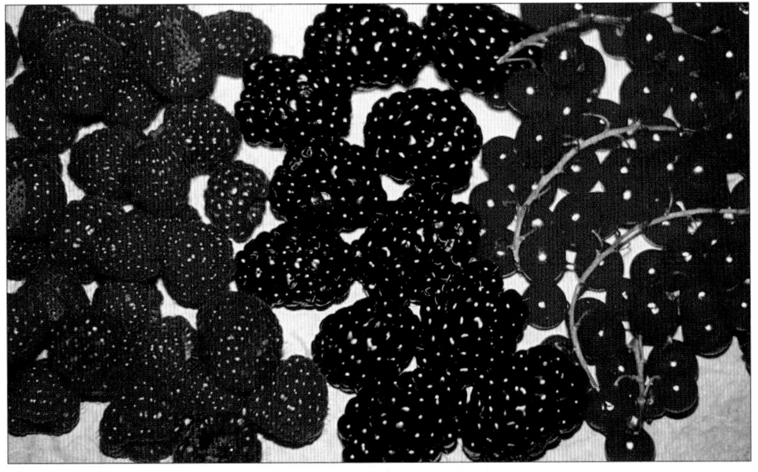

▲ From left to right: raspberries, blackberries, and red currants

Sloes

Pick the sloes after the first frost and fill the container about a third full with them with their pits intact. The sloes themselves also taste excellent afterward. The distillate has an intense note of marzipan flavor from the pits. You can also use dried sloes (see chapter 5) of course.

Snowy mespilus (or juneberry)

These bushes with small, ruby fruits grow in many gardens without the gardener even realizing how much good flavor is contained within the berries. You can use the fruit for infusing if they're soft and fully ripe. Distillates made from them are also excellent.

Spruce shoots

For this infusion, use light green, fresh, and very soft spruce shoots. You can collect them in May. Fill the container a third of the way full and leave it to infuse for as long as you want. The flavor is mildly resinous. You can also use the fresh shoots of other coniferous trees such as mountain pines, Swiss pines, or larches.

Strawberries

You should only use wild strawberries, or maybe very small berries from a garden. The strawberries that are available in supermarkets are much too watery and will contribute almost no flavor.

Fill the container a third of the way or halfway full with wild strawberries and let the whole thing sit for at least three weeks, but no more than one and a half months. The result will be a very fruity strawberry schnapps. The strawberries work very well in desserts after you use them for infusing.

Swiss pine (Zirbengeist)

Harvest the cones of a Swiss pine in early summer, while they're still soft and sappy, and cut them either in half or into quarters depending on their size. You can also use young shoots instead of pinecones (see "spruce shoots") and the infusion will still be excellent. Fill the infusion container a third full at most. After four to five weeks, the infusion will be finished, but you can also leave it longer. The infusion will be very resinous, and the schnapps won't be drinkable until you distill it. You can also use the cones of other coniferous trees.

There's a second way to make Zirbengeist, but it's very laborious. Collect the seeds from the inside of the pinecones in October, then boil them for two to three days in the alcohol under reflux (i.e., cooling the steam down and sending it back into the kettle). After that, the flavor will be intense enough that you can distill it.

Wormwood (absinthe, "the green fairy")

Absinthe used to be illegal on the grounds of its "devilish" effects. However, these effects have more to do with its alcohol content and the amount drunk than with its main ingredient, wormwood. Anise is added in different concentrations depending on the product. Essentially, absinthe is nothing more than an herbal liquor.

There is no set recipe for it; everyone makes it differently, both in terms of the ingredients used and their quantities.

Here is our version for a gallon (5-liter) batch: 2.5 ounces (70 grams) of wormwood, 0.4 ounces (10 grams) of anise, 0.2 ounces (5 grams) of rue (*Ruta graveolens*), 0.2 ounces (5 grams) of sage, 0.2 ounces (5 grams) of lemon balm, and 0.2 ounces (5 grams) of althaea leaves. Fill the infusion container a third full with the herbal mixture.

> ☺ Tip: You can also make more of the herbal mixture and freeze it. You can then thaw some whenever you need it; the flavor of freshly picked herbs keeps excellently.

Let it infuse for five to six weeks. Then you'll just have to filter the finished absinthe. If you prefer a clear schnapps or the taste is too bitter for you, you can just distill the infusion.

Another version: 1 ounce (25–30 grams) of wormwood, 2–3 ounces (50–80 grams) of anise seeds, 1.5–2 ounces (40–50 grams) of fennel, and a total of 2 ounces (50 grams) of lemon balm, peppermint, cilantro, hyssop, tansy, veronica, garden angelica, and star anise per quart (liter) of alcohol. Distill the infusion along with the herbs. If you want to produce a classic green absinthe despite the distillation, you can insert a sprig of rue into the distillate for a couple of weeks. Or, you can leave the following mixture in the distillate for a couple of days: 0.042 ounces (1.2 grams) of wormwood, 0.042 ounces (1.2 grams) of hyssop, 0.042 ounces (1.2 grams) of lemon balm, and 0.0052 ounces (0.15 grams) of peppermint. This mixture is also enough for 1 quart (1 liter).

CHAPTER 7

Spirits

The principles of spirit production

Producing spirits is very easy and has a long history stretching all the way to Paracelsus in Europe. Famous examples include Himbeergeist (raspberry spirit), gin, jenever (Dutch gin), aquavit, and ouzo. Their production is based on the following principle: the kettle contains tasteless alcohol (wine or diluted liquor like vodka, grain schnapps, or a neutral spirit like Everclear). Herbs, spices, or fruits such as raspberries, juniper berries, or anise are placed in the steam area above it. When the kettle is boiled, alcoholic steam flows through these products and takes on some of their flavor. That's why this process is also called vapor infusion. If you place the plant material not directly into the liquid but above it, you will get a clearer and finer fragrance as well as taste. Steaming the products enables you to avoid overcooking and boiling away their flavor and aroma.

Base materials

You'll need a steamer basket for this method of distillation (see chapter 3, Schmickl's stills). Its job is to hold the herbs or spices directly above the surface of the liquid so that the steam can flow through them during distillation.

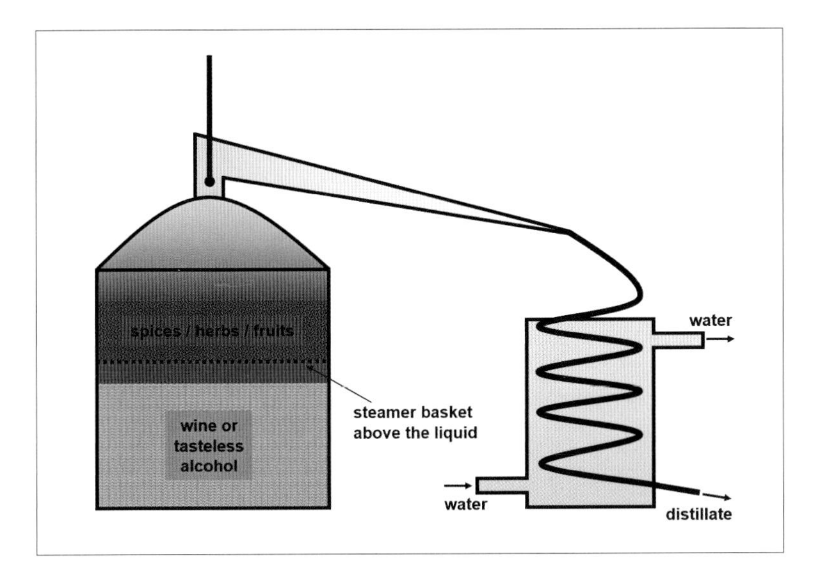

◂ Spirit production

▲ Spirit production: Pour in wine or diluted grain schnapps, place herbs or spices (e.g., anise) in the steamer basket, and distill as usual.

The "carrier" should always be an alcohol that is as taste-neutral as possible to avoid contaminating the flavors. Suitable choices include vodka, grain schnapps, neutral spirit, tasteless alcohol made with turbo yeast, and alcohol purified with activated carbon. Always dilute the alcohol to about 10–12 percent ABV.

Another cheap but excellent option is dry white wine. You don't need to worry about quality; store-brand wine from a supermarket will do just fine since you can't distinguish quality or price from the distillate. Commercially available wine doesn't contain any heads or other harmful substances, which makes the distillation especially simple. It also contains an alcohol content of about 11–12 percent ABV, meaning that it will hardly need to be diluted at all after the distillation, as this corresponds to an ABV of about 47–50 percent in the distillate. The distillate will have a sharply stinging smell, but this is not due to heads but the sulfurous acid that forms when sulfurized wine is boiled. Almost all wines are "sulfurized" (i.e., infused with sulfite salts). You certainly shouldn't be surprised if you have a headache the day after enjoying too much of such a wine. However, sulfurous acid has one big advantage: it's very volatile! That means that if you use a mixer attachment for power drills, an electric hand blender (electric mixer is not sufficient), or a milk frother to stir in air for about two minutes per quart (liter)—vigorously enough that it foams (!)—the stinging in your nose will disappear entirely and a full herbal or fruit aroma will take its place. Tails will also begin to appear as usual at 196°F (91°C).

Finally, we need the fruits or herbs that the flavor will be extracted from. Depending on what type you use, a certain amount of water will be formed from the fruits or herbs, especially if you use frozen fruit like raspberries, for example. If you use wine in the distillation, the amount of this water that forms will determine whether you'll need to dilute the distillate any further or not.

Recipes

The amounts given in the following recipes are per 1.6 quarts (1.5 liters) of taste-neutral alcohol with an ABV of about 11–13 percent. You should expect to end up with about 12 fluid ounces (300 milliliters) of distillate (diluted at 43 percent ABV). If you decide to try your hand at making your own creation or mixture, you should be cautious with your doses, especially of herbs because their flavors are very intense, and an excessively intense flavor is unpalatable. All herbs and spices can be used fresh, dried, or frozen. Dried herbs sometimes taste a little different in the spirit than fresh or frozen ones.

Absinthe

See wormwood.

Anise (ouzo, raki, Pernod, sambuca)

Place 1 ounce (30 grams) of anise in the steamer basket to make a wonderful, fine ouzo or raki. Add 0.7 ounces (20 grams) anise seed and 0.7 ounces (20 grams) fennel seed instead of pure anise for a particular mild, sweet taste. Adding water at a 1:2 ratio to the 43 percent ABV spirit before drinking will give both varieties the typical nice white color.

Pernod is an anise-based mixture of several spices, including mint, fennel, coriander, and others. Other than anise seed, sambuca contains star anise, licorice, and sugar. Thus, this beverage is actually a liqueur.

There are two different kinds of anise: star anise and anise seed. Star anise is a star-shaped legume about 0.8–1.2 inches (2–3 centimeters) in size with red seeds. For the spirit, coarsely crush four to five dried stars or break them up with tongs and place them in the steamer basket. Commercially available spirits always use star anise because it's cheaper than the higher-quality, finer anise seed. Anise seed is a grain, about 0.1–0.2 inches (3–4 millimeters) in size, that is often used in Christmas cookies. You can find them in the spice aisle of any supermarket under the label of "Anise Seed Whole."

Apples

It's hard to believe, but apple spirit's flavor almost exceeds that of apple brandy. Dice four fresh apples with yellow flesh (apples with white flesh and green skin are too little bold for this purpose) and place them in the steamer basket. If you're using dried apples, use 4.5–5 ounces (125–150 grams), coarsely chopped.

Aquavit

See caraway.

Caraway (left) is often confused with cumin (right). Both are excellent ingredients for spirits, but they have a different taste. ▸

Bananas

Dice three bananas without their peels. It's best to use overripe bananas. The "mini bananas" that are sometimes sold in supermarkets produce a more flavorful spirit than the large ones.

Barberries

These small red berries are primarily used in East Asian cooking. They produce a fruity and slightly tart taste. You'll need about 7 ounces (200 grams) of dried berries for the spirit.

Caraway (aquavit)

This spice is very intense, so you only need to use 0.35 ounces (10 grams) of it. In aquavit production, dill seeds are generally included in addition to or in place of the caraway. About 0.18 ounces (5 grams) of caraway and 0.53 ounces (15 grams) of dill seeds would be an optimal mixture. Other components in smaller amounts are cilantro, fennel, cinnamon, and cloves.

The name aquavit is used in the Scandinavian region for the described spirit. Don't confuse this appellation with the Italian aquavit, which is a kind of grappa without any spices or herbs.

Cherries

Sour cherries are the best for making spirits. Other types only contribute a small amount of flavor. Dried sour cherries have also proven to work well. Use about 8.8–10.5 ounces (250–300 grams).

Christmas spirit

To make Christmas spirit, break up a stick of cinnamon, about 0.2–0.4 ounces (5–10 grams), with some tongs and dice about 0.3 ounces (8 grams) of fresh ginger. Also add about twenty cloves and 4.4 ounces (125 grams) of chopped, dried apple rings. This mixture has a flavor so strongly reminiscent of punch stands at Christmas markets that we named it Christmas spirit.

Cinnamon

This spirit is especially good during the Christmas season. The smell alone is so intense that it puts you in a Christmassy mood. Break up one stick of cinnamon, about 0.17–0.35 ounces (5–10 grams), with some tongs for this spirit.

Cloves

Place them uncut into the steamer basket. They will not produce a great taste; they'll just make the spirit spicy. In mixtures such as Christmas spirit (page 112), however, you shouldn't discount cloves. But be sure never to add more than twenty-five of them or it will taste too spicy.

Cocoa, chocolate

Use about 1–1.5 ounces (20–40 grams) of cocoa powder, depending on your personal taste. Before placing it in the steamer basket, put a paper towel over the basket so that the powder doesn't fall to the bottom of the kettle and burn.

Drinkable cocoa mixtures can also be used, but you'll need about 7–10.5 ounces (200–300 grams) of it because they contain a relatively low proportion of cocoa. If the mixture can be dissolved in the alcohol, a basket is not necessary.

Alternatively add approximately 2 cups (200 grams) grated chocolate directly to the alcohol without a steamer basket. To guard against boil over, you should add a few drops antifoam.

Coffee

About an ounce (25 grams) of roasted coffee beans broken up with a hammer provides an exquisite coffee aroma. Since the quality of the beans has a serious impact on the taste, you should not use preground drip coffee or else the spirit will come out bitter.

Elderflowers

Place about 2.5 ounces (70 grams) of fresh or frozen blossoms along with their thin stems in the steamer basket. Like raspberries, elderflowers can be frozen easily. Be careful when you're picking them: don't wash the blossoms or you'll lose the pollen, which is the source of their flavor. As when using them in mashes or infusions, citrus fruits can make their flavor "fizzier." If you wish to do so, add about a quarter of a diced lemon and half of a chopped orange peel (including the white parts) to the steamer basket.

Fennel

For cooking, fennel certainly isn't for everyone, but it produces a refreshing spirit with an excellent flavor similar to anise. About 1 ounce (25 gram) of dried fennel seed (whole) is all you need. It blends well with other herbs such as mint, for example 0.9 ounces (25 grams) of fennel seed and 0.28 ounces (8 grams) of mint.

A particular mild, sweet-tasting ouzo is a mixture of anise and fennel seed, see details under anise.

Gentian

Gentian also makes a very intense spirit. Use about 1–2 ounces (20–50 grams) of chopped dried roots or about 3.5 ounces (100 grams) of diced fresh roots, according to taste. The advantage of making a gentian spirit and not a brandy is the time and effort: here you know after about half an hour whether the amount of gentian was enough or too much. As aforementioned, readily chopped, dried gentian root is available at distiller suppliers.

Ginger

Finely dice about 0.5 ounces (14 grams) fresh ginger root; use twice as much if you're working with dried ginger. Be sure not to add too much to the steamer basket or your spirit will be extremely spicy.

A small piece of advice for people who don't normally like ginger: its spirit is simply excellent and absolutely worth trying. The taste isn't "gingery" at all, but mild and aromatic. Ginger is also good in mixtures (see Christmas spirit).

Ginger is often added as a taste enhancer in cooking. This effect also comes through very well in spirits. Since you don't want the taste of the ginger itself to be present, about 0.07 ounces (2 grams) of a diced, fresh root is enough. Raspberry spirit made from 7 ounces (200 grams) of raspberries and 0.07 ounces (2 grams) of ginger has a more intense raspberry flavor than pure raspberry spirit, for example.

Grains

Leave about 1 ¼ cups (300 milliliters) of barley soaking in a liter of 12 percent ABV overnight and then distill with it in the steamer basket. Use about 6 cups (1.5 liters) of 12 percent ABV tasteless alcohol as your starting liquid. It's astonishing how intense the flavor of this spirit is.

Hazelnuts

Briefly roast 7–10.5 ounces (200–300 grams) of hazelnuts and then crush them finely. You'll get a very intense flavor if you distill a hazelnut infusion (see chapter 6) and place freshly roasted nuts in the steamer basket.

Herbs

There are countless possible combinations and varieties you can use. You can either use a mixture or just a single type of herb—fresh, frozen, or dried. Herbs that work particularly well include sage, wormwood, lemon balm, and althaea. Place about 1–2 ounces (30–50 grams) of fresh or frozen herbs in the steamer basket.

Juniper berries (Jenever, gin)

For a pure juniper taste like jenever (Dutch gin), the precursor of the English gin, place 1.4 ounces (40 grams) of dried, whole berries in the steamer basket. Don't use crushed juniper berries or you will get an excessively strong resinous flavor. Obviously, you can also use fresh berries, in which case you'd also need about 1.5 ounces (40 grams).

◂ Juniper branch with berries

Gin is in fact a mixture of several spices. The exact composition depends on the respective producer and is top secret, of course. The following recipe taste nearly the same as the original: 1.4 ounces (40 grams) juniper berries, a quarter finely diced lime, a quarter chopped orange peel, 0.3 ounces (8 grams) minced cucumber, 1.8 ounces (5 grams) dill seeds, 0.1 ounces (3 grams) anise seeds, 0.1 ounces (3 grams) chopped fresh lemongrass, 0.1 ounces (3 grams) crushed cinnamon, 0.07 ounces (2 grams) chopped fresh mint, 0.07 ounces (2 grams) diced fresh ginger, 3 pieces of cloves.

Lemons

Thinly peel about three to four untreated lemons (without peeling off the white part) and place the peels in the basket. Alternatively, you can use two to three whole fruits diced into small pieces. If you just use the peels, the flavor will be more intense, but it will also taste a little sharper due to their higher essential oil content.

Lemongrass

Depending on how intense of a flavor you want, 0.7–1 ounce (20–30 grams) of dried lemongrass is enough, or 1–1.8 ounces (30–50 grams) of fresh lemongrass. Don't forget to cut it up. Dried lemongrass spirit has a very slight flavor of hay.

Limes (daiquiri, mojito)

Dice the entire fruit, with its peel, into small pieces and place them in the basket. Three limes will give you an excellent fresh aroma like a daiquiri cocktail.

Add about 1 tablespoon (4 grams) of mint to make mojito spirit.

Mint, peppermint

About 0.3 ounces (8 grams) of dried mint or 0.4 ounces (10 grams) of fresh mint will provide an exquisite flavor. It mixes well with many herbs and fruits.

Some examples of ingredients that each would mix well with 0.3 ounces (8 grams) of dried mint: 0.5 ounces (15 grams) of anise seed, 0.5 ounces (15 grams) of fennel, 7 ounces (200 grams) of raspberries, one diced lime, or one diced lemon.

Nutmeg

Nutmeg has a very intense flavor, so you only need to use ½–1 teaspoon (1–2 grams), ground.

Orange (Cointreau)

Thinly peel unwaxed oranges without peeling off the white part. Three to four oranges are enough. Alternatively, you can squeeze two whole oranges (otherwise, the alcohol will be too heavily diluted) and finely dice the whole peel. This method will give you a milder flavor. To get a sweet, Cointreau-like taste, add about 0.2 ounces (5 grams) of crushed cinnamon bark.

Raspberries ▸

Ramson

Place about fifteen chopped ramson leaves in the steamer basket. Ramson spirit smells similar to garlic and, astonishingly, has a very fine, mild taste.

Raspberries

Raspberries are truly an all-purpose fruit. You can make anything with them, and their flavor always comes through strongly. Place about 7 ounces (200 grams) in the steamer basket. If you pick raspberries in the summer, you can freeze them and add them directly while still frozen later. Store-bought frozen raspberries work excellently too.

> ☺ Tip: Add 0.07 ounces (2 grams) chopped, fresh ginger as a taste enhancer (see also ginger recipe).

Rue (*grappa di rutta*)

Use 3.5 ounces (100 grams) chopped fresh or frozen rue. The flavor is the same as the Italian *grappa di rutta*, although the spirit is a clear, colorless liquid and the grappa is a green-colored infused alcohol.

Spruce shoots

Collect shoots of spruces while they're still light green and soft in May and use about 4 ounces (120 grams) of them for the spirit. You can also use frozen shoots. The flavor is astonishingly fresher, more intense, and also a little more "woody" than the distillate of the infusion (see chapter 6). You can also use the shoots of other coniferous trees (Swiss pine, larch, mountain pine, maritime pine, etc.).

Swiss pine

"Swiss pine spirit" is usually a distillate from a Swiss pine infusion (see chapter 6). However, you can attain a slightly bolder taste by directly placing five to six quartered, unripe (still soft on the inside) pinecones in the steamer basket, without infusing them first. You can also freeze the pinecone pieces and use them later. In this case, just add them to the steamer basket while they're still frozen.

Young cones of other coniferous trees (spruces, larches, mountain or maritime pines, etc.) can also be used to make spirits.

You can make the taste even "pinier" than with pinecones by using about 2.5–3.5 ounces (70–100 grams) of young shoots that you collect in the early summer.

Wormwood (Swiss absinthe)

Pure, it's best to use 1–1.4 ounces (30–40 grams) of chopped fresh or frozen wormwood.

The following mixture will give you absinthe spirit, which has nearly exact the same taste as the colorless and clear original Swiss absinthe, not to be confused with the green, sticky-sweet Bohemian absinthe liqueur. The Swiss enjoy their absinthe the same way we personally prefer: diluted to 43 percent ABV and with a bit of water eventually added. No ice, no sugar, and no lighting the absinthe while trickling it over an absinthe spoon containing a sugar cube.

Mix 0.9 ounces (25 grams) of chopped fresh or frozen wormwood, 0.5 ounces (15 grams) of anise seeds, 1 ounce (30 grams) of fennel seeds, 0.3 ounces (8 grams) of chopped fresh or frozen lemon balm, 0.03 ounces (0.8 grams) of cinnamon, 0.03 ounces (0.8 grams) of cloves, 0.03 ounces (0.8 grams) of ground nutmeg, and 3 pieces of crushed allspice. Put the mixture in the steamer basket and distill with 1.6 quarts (1.5 liters) of alcohol with 12 percent ABV.

If you want the distillate to take on the well-known green color of French or Bohemian absinthe, you can add one small stem of rue to it for every 10 fluid ounces (300 milliliters), or a mixture of 0.04 ounces (1.2 grams) of wormwood, 0.04 ounces (1.2 grams) of hyssop, 0.04 ounces (1.2 grams) of lemon balm, and 0.005 ounces (0.15 grams) of peppermint. Allow the herbs to infuse until you've reached the desired color, which generally takes a couple of days. Store it in darkness or the green color will be brown instead.

Comparison of the three distillation methods presented in this book

The following overview will summarize the differences in flavor transmission between the three distillation methods.

If you make a high-grade (20 percent ABV) mash from blossoms or fruits and also put fruits of the same type in the steamer basket during distillation, you'll get the most intense possible flavor. This method is especially useful if you're working with sour fruits or fruits without much flavor such as blackberries. Classic schnapps fruits (prune plums, pears, apricots, etc.) are distilled "normally"; the amount of flavor transmitted via this method is really completely sufficient.

If fruits are infused and the steamer basket is also used, you'll get an especially intense flavor, similar to when using a mash. This method is especially suitable for herbs or berries that you only have a limited amount of, such as sloes. The classic example of a distillate made from an infusion is nut schnapps: green, unripe walnuts infused and distilled. If you're using products without much flavor—like blackberries, huckleberries, or currants—you can add fresh fruits to the steamer basket during the distillation in order to enhance the flavor.

Method	Flavor transmission intensity
Mash + fresh fruit in the steamer basket	***
Mash	**
Infusion + fresh fruit in the steamer basket	***
Infusion	**
Spirit	*

◂ Aroma and flavor transmission with the different production methods.

Spirit production results in the lowest amount of flavor being transmitted. It's most suitable for herbs, spices, or very flavorful fruits such as raspberries. For spices like juniper, anise, caraway, cinnamon, etc., the other methods are simply too intense. But blossoms (e.g., elderflowers) can also give you a very good result. The spirit method has one decisive advantage: the base ingredients can be used directly in their original forms, regardless of whether they're fresh, frozen, or dried. You also don't need to spend weeks infusing, months fermenting and storing a mash, or years storing a brandy. If you maintain the same amount of alcohol as advised at the beginning of this chapter, it will take about half an hour to produce a delicious spirit.

CHAPTER 8

Essential Oils

Pure, all-natural essential oils, which are often used for medicinal purposes, are always made via steam distillation or cold pressing in the case of citrus fruits. You do get a higher yield if you use organic solvents such as hexane, petroleum ether, or chlorocarbons, which is why oils made this way are cheaper, but these solvents unfortunately always leave traces of themselves behind, making the oils unsuitable for medicinal use.

For the steam distillation, fill the kettle with water up to the level of the steamer basket bottom, and place the substance from which you wish to extract the oil in the steamer basket above it. If the substance is just placed in the water and boiled, you will get a very low yield or possibly nothing at all because the steam, not the boiling water, can carry away the essential oil, which has a higher boiling point than water. Aside from that, the products should be steamed instead of boiled to avoid overcooking the damageable compounds of the essential oils. Spirit production (see chapter 7) is also a steam distillation, but in that case the organic solvent ethanol (alcohol) is used in place of water.

Since oil and water cannot mix, the essential oil will become separated and float on the top, with a few exceptions. You should therefore let the distillate run directly into a narrow-necked bottle. Once the bottle is completely full, you just have to suck up the oil from the bottle neck using a syringe or a pipette. Some oils settle at the bottom of the bottle, such as cinnamon oil or clove oil.

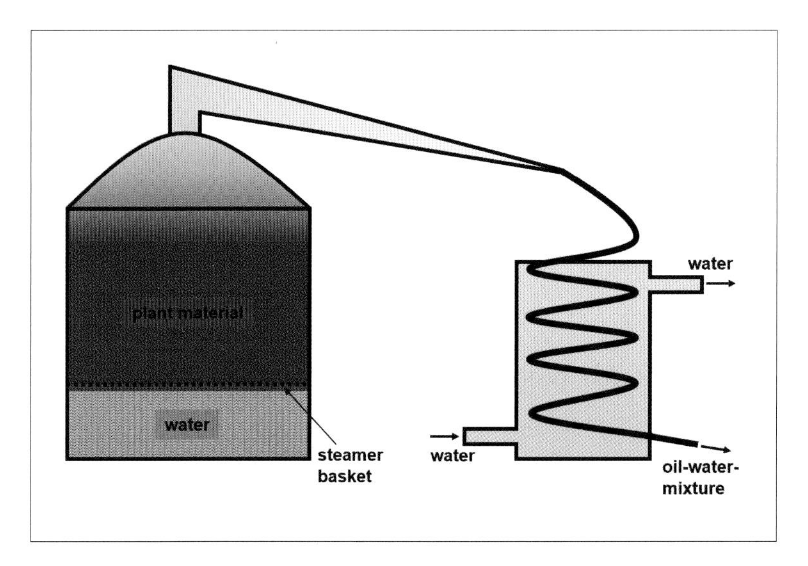

◂ Apparatus for producing essential oils

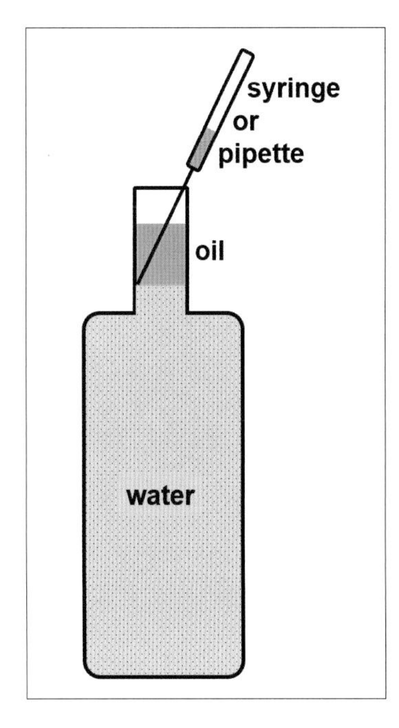

▲ Separating the essential oils from the water

Which products can you use to create essential oils? It's best to take a look at the selection of essential oils available in drug stores, pharmacies, or other suppliers to see the seemingly endless variety. Some examples: rose petals, mountain pine, fir cones, oranges, lemons, grapefruits, cinnamon, and peppermint.

Finely mince the substance you want to extract the oil from and add as much of it to the steamer basket as possible. If you're using pinecones, cut them up into small pieces first. If you're using citrus fruits, just use cut-up pieces of the peel. For approximately every gallon (4 liters) of plant material and 1.5 to 2 quarts (1.5 to 2 liters) of water, depending on the type of plant, you can expect to make about 0.07–0.7 fluid ounces (2–20 milliliters) of highly pure essential oil.

Our book *The Essential Oil Maker's Handbook* describes this process in detail as well as methods of creating essential oils aside from steam distillation. It also includes tips on preparation, harvest times, and information on working with over a hundred domestic and exotic plants. The last part of the book gives over forty core recipes for how to use your essential oils to create a variety of products, from bath additives to facial and body care products to perfume.

CHAPTER 9

Drinking Culture

Bottling and labeling

Now that your schnapps has been created through your painstaking labors, it deserves the proper packaging, too. Even the best schnapps cannot come into its own in a beer bottle with its original label crossed out. Besides which, visuals play a significant role in enjoying the beverage. You should therefore take heed of the following points:

Glass bottles

Always bottle schnapps in transparent, white glass bottles. In any other sort of bottle, you can't see that you have a crystal clear liquid without cloudiness or tinges of color. You should avoid green or brown bottles. You don't need to buy the expensive triangular bottles that are so trendy right now; regular bottles will do just fine.

An example: You're surely familiar with the small, white, piccolo champagne and sparkling wine bottles that contain 7 fluid ounces (0.2 liters) and have easy to remove (warm water and dish detergent) labels. Go enjoy a nice sparkling wine, and at the same time you will have acquired a nice schnapps bottle.

Cork stoppers, metal screw tops

Your schnapps should be sealed as well as possible. The piccolo champagne bottles mentioned above have metal screw tops, but cork stoppers work just as well. Be sure not to use corks or screw tops made of plastic. They not only look ugly, toxic substances can also sometimes be dissolved from them.

▲ Stylish glasses

◂ Cork stoppers, metal screw tops, and heat-shrinkable caps

Heat-shrinkable caps

Nice, colorful, heat-shrinkable caps, which are available in many colors and sizes at co-ops and specialty winemaking shops, give the whole thing an excellent finishing touch. Just heat it with a hot air gun from the hardware store from the top to the bottom, and your bottle will look wonderful.

Labels

Be sure to label your bottles or you'll eventually end up mixing them up. However, you shouldn't just use normal white household labels and write on them with a pen, as this looks unprofessional. You should at the very least use commercially available preprinted labels. They're certainly not the best solution, but they're better than the other approach.

Of course, you can make the very best labels yourself with your computer. There are excellent self-adhesive sheets for inkjet and laser printers, in standard letter format or as pre-cut labels. Just make a simple template using your computer and decorate it with pictures of fruits. This method also helps you create very nice personalized gifts for specific people or celebrations.

Some examples of labels ▸

Checking the taste and quality

Now that your work is complete, you should also enjoy the fine product according to all of the rules of the art. Keep the following points in mind while doing so:

- Use a tulip glass. This is the only way for the flavor to develop properly.
- Drink the distillate at a temperature of 64°F–68°F (18°C–20°C).

If you're offered a distillate, you can tell whether it's of high quality based on the following points:

- Schnapps is usually served ice cold in restaurants. This doesn't allow you to perceive any of its flavor or smell. Premium distillates are never served this way because the goal is to impress you with its flavor.

◂ Tulip glass

- If you can recognize the smell of glue, the heads separation was not carried out properly. In this case, you can expect a headache the next day.
- Especially in rural areas, distillates with an ABV greater than 50 percent are often offered. Aside from a burning and stinging in your nose, you won't be able to perceive anything because the distillate is simply too strong. However, that says nothing about its quality, as high alcohol contents always lead to this sort of irritation.
- You can recognize a good schnapps because it will exhibit lots of body and vigor on the palate and in its aftertaste. The taste should still be recognizable on the palate for a while afterward, too.
- Smell the empty glass. If it was a high-quality distillate, you should still be able to perceive its scent for a while in the empty glass. If the scent is volatilized after a few seconds, the drink was artificially flavored. A simple test shows the same: apply one drop on a sugar cube, put it in your mouth, and suck on it. The flavor should last more than a few seconds.

Other countries, other customs

▲ Drinking culture on the Mekong

Regardless of which country you find yourself in, alcohol always seems to be an important element of culture. In Laos, for example, every village has its own distiller that produces the famous *lao-khao*. This rice brandy is unrelated to Chinese rice schnapps—it's a much higher quality product. Quality obviously varies from village to village, but based on the materials used and the means of the people, all you can do is marvel at and enjoy the high quality. We found one particularly extravagant still on the banks of the Mekong River. They had repurposed an old oil barrel for use as a still. A wok was used as the condenser, and a bamboo cane led the condensed schnapps out of the side into an earthenware jug. Cooling water was manually circulated with an old U.S. GI helmet! The mash was a mixture of water, sticky rice, and dried and ground manioc roots, and a mysterious remedy, probably a kind of yeast.

Still on the Mekong in Laos ▶

The mixture was poured in solid, thick pottery jars, which were shut airtight with pottery caps and rags tightened with thick ropes. Finally, the jars were buried in the cool sand of the Mekong banks and dug up again for distilling after about a week or two.

▲ Myanmar, Bagan: Distilling palm wine

There's another very nice custom in Laos. The fresh fermenting rice wine—similar to German *Federweißer*, or "feather white" (in Austria it's called *Sturm*, which means storm and better describes what happens: the storm caused by the heavily fermenting wine in the glass and the storm in your bowels after drinking this digestive stimulant)—is poured into an earthenware jug with a wide, flat rim up to the very top. Then a long bamboo cane is used as a drinking straw. The host holds a drinking glass full of water in his hand, about a cup (a quarter of a liter), and pours it fairly quickly into the jug, which is filled to the brim. To avoid losing his honor, the guest must drink quickly enough through the bamboo cane so that no liquid spills over.

In Myanmar, driving from Bagan to Mount Popa, our driver suddenly decided to stop a few minutes after leaving Bagan. Without request, he brought us to a village distillery! The principle of the still was the same as what we had become acquainted with in Laos, but even a bit more archaic: the kettle was made of an earthenware jar, the only "modern" utensils were an iron wok used as the condenser and two old, used, glass bottles utilized as receiver. The distilling liquid was coconut palm wine, locally called *htan yay*. To tap the palms, the tips of young coconut flowers at the top of the tree were cut and the outflowing palm sap was collected in an earthenware jar. After fermentation in open earthenware jars, without adding any additives or mysterious remedy, the palm wine of about 4 percent ABV was distilled once. We didn't observe whether heads were separated or not. The resulting distillate had an extraordinary pungent flavor, like a mixture of undiluted heads, bad-tasting grappa, and sweaty feet (butyric acid!), all out of proportion to the smooth and mild *lao-khao*.

▲ Myanmar, Bagan: Palm sap is collected at the top of the tree

We had a very different experience in Honduras in Central America. It's almost impossible to find good schnapps there; it's much more common for the local Garifuna people to distill a "hell water," which is the base for the Guifiti, an infused alcohol with herbs and roots. When we told the manager of our hotel on the Caribbean island of Guanaja that distilling schnapps was our hobby, he perked up and was very interested to learn how to make good schnapps. No sooner said than done. We built a still out of his kitchen utensils. The lime spirit we made out of half a gallon (2

▲ Kitchen Still in Guanaja

▲ The receiver is located inside of the still

liters) of white rum, which we diluted before distillation, and 6.5 pounds (3 kilograms) of lime wedges was simply delectable!

The high quality of Scotch whisky is of course well known. The Scots are masters of malt and distilling to the finest detail. The whiskies from the Scottish islands, such as the island of Islay, have an especially smoky flavor. The flavor that comes from roasting the barley with peat usually takes some getting used to, but then it's simply excellent.

To wrap things up, let's take another quick look at the Caribbean. Its rum is very famous, of course. The white rum tastes like "firewater" and is only suitable for cocktails and mixed drinks or for flambéing. The dark rum, on the other hand, is stored in barrels for years like whisky, after which time it is certainly one of the finest spirits in the world. A brief word on Bacardi rum: its taste is very different from that of every "classic" type of dark rum because the Bacardi family has added a secret herbal mixture ever since the nineteenth century.

Rum storage tanks in the Caribbean ▸

CHAPTER 10

Legal Situation

The laws relating to distilling vary greatly throughout Europe. In the southern countries, the laws are more casual—in some countries, in fact, there are no restrictions on distilling at all and you can distill as much and as often as you want to for your own private consumption. The situation is different in Germany, Austria, and Switzerland, however. The government strictly controls the number and location of stills here.

In Germany and Austria, the legal situation also depends on whether you're dealing with a settlement still or a licensed still.

In licensed stills, the alcohol is checked (after it's distilled). After the condenser, these stills have a permanently attached measurement point where the alcohol content and flow rate are checked. This allows customs officials to determine how much alcohol with what alcohol content is being produced, and thereby calculate how much is due in alcohol taxes. Because the measuring apparatus is very expensive, licensed stills are generally only found in large distilleries.

The alternative is a settlement still. These are mostly relevant for small-scale distillers. When using these stills, the local customs agency must be informed of what type and amount of mash is being distilled, and they use that information to determine how much alcohol tax needs to be paid.

The difference thus lies in the fact that in licensed stills, it's the end product that is checked, so it doesn't matter what was distilled in the kettle earlier. For settlement stills, on the other hand, it's the starting product (the mash) that is used in the calculations. This means that no alcohol-forming substances (sugar, for example) can be added to the mashes in this case because that would invalidate the foundation of the calculations. Sugar can only be added to mashes in licensed stills.

Germany

In Germany, anyone has the right to obtain a license to operate a licensed distillery. There are distilling licenses for fruit, wine, grains, potatoes, sugarcane, and many other things.

There is a hard limit of about 30,000 settlement still licenses, most of which are issued in southern Germany (Bavaria). If someone gives up his settlement still license, someone else in the same administrative district can acquire it. New, additional licenses are rarely issued nowadays. So-called raw material owners are people who can make their own mashes. They are also concentrated in the south of Germany and can give their mashes to someone with a settlement still license to distill. Finally, there are the cooperative fruit distillers, who work with licensed distilleries. They collect fruit from many different fruit farmers, make it into a mash, and distill it.

All stills (including homemade ones) must be registered with the customs authorities, even if they're not being used to distill alcohol. German still vendors must also notify the customs authorities of all sales of stills, along with the customer's personal information. It's illegal to advertise or sell any device in Germany that's intended for non-commercial production or purification of small amounts of brandy (§46, Abs. 1 BranntwMonG). Stills with a kettle volume of up to 17 ounces (0.5 liters) are exempted from this ban and the registration requirements as long as they are not used for commercial purposes. You should refer to §34 BranntwMonG on the subject of building your own still with a kettle volume of greater than 17 ounces (0.5 liters) or purchasing one: "Die Erzeugung von Brennereien ... gilt als innerhalb des Brennrechts hergestellt, wenn sie in einem Betriebsjahr zehn Hektoliter Weingeist nicht übersteigt." ["The creation of a still is considered legal within an existing distilling permission if its production does not exceed ten hectoliters of alcohol in one year."]

You can make schnapps for personal consumption just fine with small stills—as long as they're constructed properly. Working with small quantities has a serious advantage: after the distillation, you aren't stuck with 26 gallons (100 liters) of prune plum schnapps (that you're not allowed to sell), but a wide variety of different distillates.

Austria

Distilling licenses

Austrian laws are a little more liberal than their German equivalents. Anyone may operate a licensed still here, and it's also possible to run a settlement still, which is very common among private citizens and farmers because anyone who has access to fruit trees is allowed to apply for a settlement still license. The type and quantity of mash must be reported to the local customs authorities as described five business days before the distillation at the latest. There are only laws protecting monopolies for sugar beets, turnips, potatoes, and grains: it's illegal to distill these products with settlement stills, except for mountain farmers who have limited access to fruit.

Like in Germany, all stills must be registered, and there's a similar double registration requirement (by the seller and the buyer of a still). However, stills of up to 0.52 gallons (two liters) do not have to be registered in Austria. Nevertheless, you always have to pay the alcohol taxes if you're distilling untaxed alcohol.

You can find more on this in the alcohol tax laws, §20 and §85.

Austrian Food Code

There are very exact legal guidelines related to production and contents for all alcoholic beverages that are sold commercially. In Austria, distillates must be labeled with what class of product they are in an unmistakable fashion. The most important two are (fruit) "brandies" and the general term "spirituous beverages." You can also include whatever made-up promotional name you want, like "finest plum schnapps," but "spirituous beverage" must also be written in small letters somewhere or on the back of the bottle.

Only distillates produced according to the strict standards of the Food Code (without artificial flavors, etc.) are allowed to be labeled as "brandies" (e.g., prune plum brandy, pear brandy, apple brandy, etc.). A good brandy comes at a price and is not mass-produced.

Switzerland

Swiss law distinguishes between two types of distilling: commercial and contract. Commercial distillers produce and sell spirits, whereas contract distillers just distill under contract. They also distinguish household distillers. These are the only private citizens who are allowed to distill, under the condition that they are engaged in farming. Household distillers can distill for their personal consumption without having to pay any taxes. Other private citizens can bring a mash to a contract distiller (similar to the situation in Germany).

All stills, regardless of the size of the kettle, must be registered; there is no minimum volume for this requirement. Interestingly, however, stills with a kettle volume of up to 0.79 gallons (three liters) can be purchased without any registration, as long as they are only used to produce essential oils (through steam distillation).

Spirit productions and distilling infusions*

If you use store-bought (i.e., already taxed) distillates for spirit production or infusing, there is no further tax obligation after the distillation. You still must register your plans to do so with the responsible customs authorities, however.

*Applies in Germany, Austria, and Switzerland

If you do so, be careful that what you're buying is really a spirituous beverage (i.e., a distillate). Wine has an excessively low tax class. Before distillation, you're allowed to dilute a high-percentage liquor to wine strength, for example. Interestingly, distilling spirituous beverages that have already been taxed falls under the category of "alcohol purification."

The United States

The laws surrounding alcohol distillation in the United States are fairly prohibitive, although they are not as different from some European regulations as they may seem at first glance. Many differences also exist in the laws between individual states.

Americans may not produce spirits for beverage purposes without paying taxes and without prior approval of paperwork to operate a distilled spirits plant. According to 26 U.S.C. Section 5171, operations as a distiller, warehouseman, or processor may be performed only on the bonded premises of a qualified distilled spirits plant. To qualify as such a plant, a registration, application for permit, and bond must be filed in addition to other supporting organizational documents.

Americans may own a still as long as it is no larger than 1 gallon. These stills may only be used for water purification or the extraction of essential oils from plants. U.S. dealers/manufacturers of stills are required to provide the address or other information of any customer who buys a still to the Bureau of Alcohol, Tobacco, Firearms, and Explosives when they request it (no warrant is required).

U.S. readers may find the following questions, from the FAQ of the U.S. Alcohol and Tobacco Tax Trade Bureau website, particularly enlightening.

> **I've seen ads for home distilling equipment in catalogs. Is it legal to buy and use a still like that?**
> Under Federal rules administered by TTB, it depends on how you use the still. You may not produce alcohol with these stills unless you qualify as a distilled spirits plant (see earlier question). However, owning a small still and using it for other purposes is allowed. You should also check with your State and local authorities—their rules may differ.
>
> A still is defined as apparatus capable of being used to separate ethyl alcohol from a mixture that contains alcohol. Small stills (with a cubic distilling capacity of a gallon or less) that are used for laboratory purposes or for distilling water or other non-alcoholic materials are

exempt from our rules. If you buy a small still and use it to distill water or extract essential oils by steam or water extraction methods, you are not subject to TTB requirements. If you produce essential oils by a solvent method and you get alcohol as a by-product of your process, we consider that distilling. Even though you are using and recovering purchased alcohol, you are separating the alcohol from a mixture—distilling.

May I produce spirits for my personal or family use?
You may not produce spirits for beverage purposes without paying taxes and without prior approval of paperwork to operate a distilled spirits plant. [See 26 U.S.C. 5601 & 5602 for some of the criminal penalties.] There are numerous requirements that must be met that also make it impractical to produce spirits for personal or beverage use. Some of these requirements are filing an extensive application, filing a bond, providing adequate equipment to measure spirits, providing suitable tanks and pipelines, providing a separate building (other than a dwelling) and maintaining detailed records, and filing reports. All of these requirements are listed in 27 CFR Part 19.

Spirits may be produced for non-beverage purposes for fuel use only without payment of tax, but you also must file an application, receive TTB's approval, and follow requirements, such as construction, use, records and reports.

More information can be found at the Tax and Trade Bureau website at www.ttb.gov/spirits/faq.shtml.

CHAPTER 11

Frequently Asked Questions

Because theory and practice famously don't always agree, we've selected the following typical questions relating to practical distilling that we've come across over the years. You can find the complete and current list on our website, www.schnapsbrennen.at.

Distillate with a slight yellowish taint

We've been distilling schnapps for a long time. Recently, we've often had the issue of the distillate not being completely pure and colorless, but tainted a light yellow. What might be causing this?

The distillate is actually always colorless. If you're talking about distillation methods that have always resulted in a clear distillate until now, I can only imagine one possible cause: something spilled over into the condenser during the distillation. Clean the still and try distilling again. If you're distilling schnapps in a way that's never resulted in a pure distillate in the past as "proof," it's also possible that essential oils from some of the raw materials you're using are responsible.

Finally, poorly welded joints in your still may be responsible for the distillate's color.

Methanol

When I buy wine at a store and then distill it, I don't need to be afraid of methanol, do I?

You're correct. Methanol is only present in trace amounts, and you also consume this much of it whenever you drink wine (see chapter 4, "heads" section).

Milky orange spirit

A question relating to orange spirit. I infused orange peels in 45 percent ABV schnapps and distilled the result. The distillate tasted very good. When I added distilled water, however, the schnapps immediately took on a milky cloudiness. I then diluted it further to 43 percent ABV. Can you give me some advice on how to make the schnapps clear again, or what I else I can do with it?

If you infuse oranges and want to drink the infusion directly, you should definitely not dilute it afterward, as doing so will always make it cloudy. A 45 percent ABV schnapps is fine for an infusion and doesn't need to be diluted any more, and not diluting will allow it to remain clear. If you want to dilute it afterward, you will experience some cloudiness due to the essential oils in the orange. You can also filter the schnapps using a pre-folded paper filter (fluted filter) grade 1 after letting it sit for fourteen days. It sometimes also helps to use two coffee filters with absorbent cotton in between as a filter and to leave the schnapps in the freezer overnight before filtering it.

Steamer basket

How does the steamer basket work in your still?

The steamer basket is a sort of metal sieve with feet. Put it into the kettle. It ensures that only the liquid part of the mash will reach the bottom of the kettle, and a filtered liquid cannot burn. The solid components of the mash will remain in the sieve. This makes it unnecessary to filter the mash to keep it from burning. Filtering before the distillation leads to a loss of flavor because much of it comes from the fruits themselves. The steamer basket can also be used to hold ingredients. To do so, place the brass feet on the kettle. Now the basket is located in the steam area of the kettle, a short ways directly above the level of the liquid. You can place herbs or spices (e.g., juniper berries for gin or raspberries) in it and just fill the kettle with cheap white wine or tasteless alcohol. Your result will be a wonderful raspberry or herbal spirit.

Fermentation lock and heads in infusions

Why is it that I only need to firmly seal the container during an infusion and don't need a fermentation lock? Will toxic heads appear if I distill the liquid?

You don't need a fermentation lock for so-called infusions because no fermentation takes place, which is the only way to form CO_2. The vodka simply extracts the flavor from the raspberries; no gas is formed (as it would in a mash), so you only need to seal the container. If you're making an infusion from store-bought vodka, you won't encounter any toxic heads because store-bought vodka doesn't contain heads.

Denatured alcohol

You write that activated carbon can remove all taste and smell substances. Can it also remove the substances that denature alcohol? As far as I know, the starting product for denatured alcohol is just normal, drinkable alcohol. Can you also get infused schnapps from denatured alcohol using activated carbon?

Unfortunately, denatured alcohol cannot be purified with activated carbon. The reason: the alcohol (you're correct that it's normal "drinkable alcohol") and the added substances behave so similarly from a chemical point of view that it's not so easy to separate them again (the lawmakers already thought of that idea). Activated carbon works on the principle of adsorption: the taste and smell substances are very large molecules compared to the alcohol. They are then held in the fine pores of the carbon, but the alcohol is not (see page 76).

Diluting

If you're making Himbeergeist (a German schnapps distilled from neutral spirit and raspberries) or another fruit brandy, should you store it for a few weeks first and then dilute it to drinking strength? Or is it better to dilute it immediately after the distillation? In my experience, the brandies don't fully develop until after a month.

If you're making a spirit, it's best to dilute it right after the distillation (as with conventional schnapps distilling). Then you can store it however you like, for example first letting it sit with the cork just set on the top for two to three weeks before sealing it so that the aroma can develop properly.

Mashes

I have two questions on the subject of mashes:

1. *Does it have a negative effect on the quality (e.g., the taste) if I freeze berries that I can't use immediately and then thaw them a few days later and make them into a mash?*
2. *Is there any disadvantage to juicing the berries and then fermenting the juice?*

1. You won't lose any flavor by freezing the berries.
2. Yes, this carries a serious disadvantage: the majority of a fruit's flavor is contained in its skin. If you only use the juice to make your mash, you'll lose much of the flavor. This is exactly why it's a good idea to distill your mash unfiltered with a steamer basket, as this method maintains all of the flavor.

Reflux distilling

A question about reflux distilling. The repeated distillations in the column naturally result in a higher alcohol content (as in a pot still with a rectifying device). But how does it affect the taste? Is it really any better than a turbo mash distilled via simple distillation?

The result of reflux distilling is a very high-percentage and pure alcohol. The principle behind it is that multiple "normal" distillations take place one after another in a device (the reflux column). It's therefore not "double distilled," but distilled five, ten, or twenty times, depending on the height and the separating power of the column. However, this strong separating effect also separates the flavor. You should only use reflux distillation to separate or purify mixtures of different liquids. It's not good to perform this kind of intentional separation when you're making schnapps because it often causes a significant loss of flavor. In extreme cases, you may even get a completely tasteless alcohol. The best method: Make a high-grade mash (with an ABV of at least 16 percent; the high alcohol content also increases the mash's flavor because it leads to much better extraction) and distill it once with a "normal" still. Because of the mash's high alcohol content, this will already produce a product with an ABV of 53 percent or more, depending on the exact alcohol content of the mash.

Activated carbon

I diluted a schnapps made of sugar and turbo yeast from 87 percent ABV to 50 percent ABV and used activated carbon with it. I planned to distill it again along with the carbon in order to attain a completely neutral flavor. Now, however, a passage from another book has confused me. Its authors claim that it's a mistake to leave the schnapps on the carbon for more than two days. The same goes for distilling with the carbon because the harmful taste particles would be freed again by this process. I'd appreciate some advice based on your experience.

There are many sources that make this claim. However, I can say the following based on my knowledge of chemistry and my extensive experience with activated carbon: the longer the alcohol is in contact with the activated carbon, the more thoroughly the smell substances will be removed. You have to imagine the carbon as a sponge, with many small pores. The taste particles are taken up (adsorbed) by the carbon. The longer the carbon remains in contact with the liquid, the more it can take up because the particles are diffused deeper and deeper into the "sponge" over time. These taste particles are bonded to the carbon via the van der Waals force. They don't just release the particles after two days; the forces are stronger than that. The same goes for the distillation. The temperature, 212°F (100°C) at most, is not sufficient to separate the taste particles from the carbon.

You would need a so-called bake-out at a temperature of several hundred degrees to separate the particles from the carbon—you'll never get them out otherwise. That would theoretically be one way to at least partially reactivate the activated carbon: just put it in the oven. In short, it is safe to distill with carbon and to leave the schnapps on the carbon for longer than two days.

Dome-shaped lids, juices, methanol, temperature curves, and hearts

1. *Does the still necessarily have to have a dome-shaped lid? And what is its function? I want to heat the mash in a water bath.*
2. *Can I use concentrated juices (apple, pear, cherry) for the mash, or do mashed pieces of fruit absolutely have to be in it?*
3. *Whenever I talk about distilling with my friends, they always ask about toxic methanol. Does it not pose a potential health risk? If I keep the temperature at a constant 148.5°F (64.7°C) for a while, that must get rid of the methanol, right?*
4. *Do you have to go through the temperature curve during the distillation or can the temperature be left at a constant 173.3°F (78.5°C), similar to the situation in my third question, to get the hearts?*
5. *How often do I need to distill the hearts; is once enough?*

1. No, a dome-shaped lid is not necessary; its function is just visual. A flat lid will work fine for a still. Why do you want to use a water bath? The only advantage to water baths is that they don't burn anything. But if the water bath is not pressurized,

you'll never reach the necessary temperature to carry out the distillation through to the tails. Use a steamer basket instead, which is a type of metal sieve that doesn't let the solid parts of the mash reach the bottom of the kettle, just the liquid.

2. You can use juices too, of course. But you should keep in mind that a very large amount of the flavor is contained in the skins of the fruits, so a mash made from juices will never be like a mash made from fruit. However, you can create very good products from juices, too.
3. The toxic methanol appears during the fermentation. If you work under clean conditions (no twigs/leaves/stems, no rotten fruit) and use a fermentation lock and cultured yeast, almost no methanol will develop. Methanol unfortunately cannot be separated out through distillation alone, so be sure to work under clean conditions so that you won't have any issues.
4. No, you must always go through the entire temperature curve. You cannot keep the temperature constant; doing so during distillation isn't even physically possible. First, increase the heat quickly to a steam temperature of 158°F (70°C) then reduce the heating power somewhat so that nothing boils over. As long as the temperature gauge is increasing rapidly, you're producing heads. If the reading remains nearly constant and only moves slowly, you're producing hearts. You'll always start to see tails at 195.8°F (91°C).
5. If you're distilling something with an alcohol content of at least 10 percent ABV, then you only need to perform a single distillation. If the alcohol content is less than that, you'll have to perform a double distillation. Double distilling is only done in order to ensure an alcohol content of 45 percent ABV in the distillate; this is the only reason. Every distillation takes away from the flavor.

CHAPTER 12

Harvest Calendar

The following calendar is suitable for European and North American continental climate and the adjacent regions of surrounding climatic zones.

	Jan.	Feb.	Mar.	Apr.	May	Jun.	Jul.	Aug.	Sep.	Oct.	Nov.	Dec.
Apples								■	■	■		
Apricots							■					
Blackberries						■	■					
Bullaces, damsons								■	■	■		
Cherries					■	■						
Currants, red and black						■	■	■				
Elderberries, black								■	■			
Elderflowers						■	■					
Gentian								■	■			
Grapes								■	■	■		
Jerusalem artichokes							■	■	■			
Medlars										■	■	■
Peaches							■	■				
Pears								■	■	■		
Prune plums								■	■	■		
Quinces									■	■		
Raspberries						■	■					
Rhubarb						■	■					
Rose hips									■	■	■	■
Rowanberries									■	■		
Sloes										■	■	■
Spruce shoots					■	■						
Strawberries					■	■	■	■				
Sweet potatoes							■	■	■			
Swiss pine						■	■					
Walnuts					■	■						
Wild plums								■	■	■		

Afterword

At least once per month, we host distilling seminars where participants have the chance to obtain a personal, active understanding of the most important steps of the brandy and spirit production process. We provide all the necessary materials and measurement devices. Each participant performs the following activities by himself over the course of the seminar:

- Determining alcohol content in mashes and distillates
- Setting up a pot still
- Distilling several different brandies
- Distilling several different spirits
- Separation of heads, hearts, and tails
- Diluting distillates

Taste test bottles are included, and each participant gets to take his own distilled brandy or spirit home with him.

Although the main language during the workshops is German, is it also possible for English-speaking participants to follow the context due to the structure of the course.

Apart from that, it is possible to lecture single persons or small groups in English at any time; please contact us for more information.

Are you interested but unable to visit us in beautiful Carinthia? Then attend our online seminars at www.distilling-fermenting-seminars.com!

We do also offer the necessary equipment and materials as described in this book; see www.schnapsbrennen.at.

Contact address:
Dr. Bettina Malle and Dr. Helge Schmickl
Ehrentalerstr. 39
9020 Klagenfurt
Austria/European Union
Tel./Fax: 0043-(0)463-437786
E-mail: schmickl@schnapsbrennen.at
Homepage: www.schnapsbrennen.at

Index

About the Authors

Bettina Malle and **Helge Schmickl** graduated from the Vienna University of Technology in 1991 with a master of science in chemical engineering and received doctorates in technical sciences in 1993. Each earned a bachelor of business administration degree from the Graduate School of Business Administration Zurich while designing, engineering, and commissioning industrial plants and managing research and development projects. Thereafter they worked as technical and business consultants until 1998.

Bettina Malle ▾

Malle and Schmickl believe it should be possible for everyone to produce exquisite spirits with fruit and herbs right from the garden. In 1998, they developed their first still, designed to maximize the flavor of alcoholic distillates. That same year, they launched their first webpage and online store and began to host small-scale distilling workshops. They published the results and conclusions of their experiments as well as detailed instructions and many recipes for crafting distilled spirits in their 2003 book *Schnapsbrennen als Hobby* (*The Artisan's Guide to Crafting Distilled Spirits*). It became the standard reference book for home distillers in German-spoken countries and is now in its tenth edition.

In the meantime they designed and constructed an optimized still for producing essential oils and hydrosols on a small scale. Due to its special designed condenser and adapted shape, the still makes it possible to obtain plenty of oil and intense hydrosols with only small amounts of base material. The two have hosted essential oil workshops and an online store since 2002 and published the reference book *Ätherische Öle selbst herstellen* (*The Essential Oils Maker's Handbook*) in 2005. The book is now in its sixth edition, and Malle and Schmickl's still is the preferred device among small-scale users in the fields of phyto-aromatherapy and herbology throughout the German-speaking region. It is widely used in many other courses, workshops, and seminaries as well as in research institutes, universities, colleges, and other educational services.

▲ Helge Schmickl

In 2008 Schmickl and Malle engineered and constructed a small-scale, modified vinegar generator that is a fixed-bed reactor with immobilized vinegar bacteria. The construction enables the use of miscellaneous packing materials, especially fruit, herbs, and spices, to improve or change the taste of the resulting vinegar. Additionally they developed a method to analyze the residual alcohol content in vinegar, which is simple, cheap, and accurate enough to determine a concentration of 0.1 percent. Their book *Essig herstellen als Hobby* (*The Artisanal Vinegar Maker's Handbook*) was published in 2010, and they have hosted vinegar workshops as well as their vinegar webpage and online store since 2011.

Bettina and Helge have been married since 2002. That same year they moved to Klagenfurt, Carinthia (Austria), where they host their seminars and conduct research and developments in fermenting and distilling. They have two children.

Also by the authors...

The Artisanal Vinegar Maker's Handbook

by Bettina Malle and Helge Schmickl; translated by Paul Lehmann

Vinegar making is a very ancient craft. Mankind first harnessed the creation of vinegar, along with its preservative and medicinal qualities, more than ten thousand years ago. Nowadays, however, most guides to making your own vinegar are limited to allowing wine to ferment on its own, often with less-than-stellar results. Truly high-quality vinegar production is an art and science in itself. Austrian distillers Helge Schmickl and Bettina Malle use their experience and scientific background to provide special insight into the creation of artisanal vinegars. Detailed, step-by-step instructions for over a hundred recipes illuminate this fascinating process for beginners, and even experienced vinegar crafters are bound to refine their techniques. *ISBN 978-1-943015-02-3. Hardcover, 192 pages.*

The Essential Oil Maker's Handbook

by Bettina Malle and Helge Schmickl; translated by Paul Lehmann

Essential oils are more in demand now than ever, but modern production methods and unscrupulous labeling practices make it difficult for consumers to know whether an oil is genuine or artificial. Healers utilizing these age-old materials now have the guidance needed to make their own essential oils and thereby be certain of integrity and efficacy. Producing essential oils and hydrosols in small quantities is easier than imagined with the guidance of Austrian master distillers Helge Schmickl and Bettina Malle. Translated from its original German, *The Essential Oil Maker's Handbook* has been revised and updated to include information on hydrosols, the aromatic water once considered a mere by-product but now recognized as a valuable substance in itself. *ISBN 978-1-943015-00-9. Hardcover, 160 pages.*

www.spikehornpress.com